Introduction

"A good picture book to trigger discussions between the Parent and the Scout."

Why read this book?
You want a fast Pinewood Derby car? You want to know *why* you have a fast car? Then this picture book is for you. Do you think you'll grow up, or already have grown up, and will want to be a scientist or engineer, or will want to at least understand some basic features of cars like air drag, bearings, wheelbase, and horsepower?

Are you looking for ways to relate a Pinewood Derby car to real people cars, so that the whole Pinewood Derby experience helps down the line when thinking about people cars? Cars coasting in neutral are very similar to PD cars, or rolling down hill, except regular cars can turn the wheels.

The audience is from 8 years old to 108 years old. The pictures and graphics are engaging for any age.

Pinewood Derby cars allow kids to use wood, hammers, nails, and creativity and competition. The kids can build the car themselves, and learn to use their hands. The scout leaders and parents themselves, by helping the scout, may learn more than the scouts from doing the activity.

This picture book compares the Pinewood Derby cars to real cars, and could be read by both the parent and the kid together, to trigger and answer questions. The emphasis of this book is part description of some of the physical reasons a PD car goes fast, reducing drag, and is part comparison of PD cars to people cars. This book is good in combination with other Pinewood Derby books that more directly describe the step by step process to making a good Pinewood Derby car.

This book has the goal of applying engineering ideas to something that most people are already familiar with, the Pinewood Derby. Maybe people will get some easy appreciation of the engineering ideas, such as energy and drag, as well as get some techniques to increase (or keep) the speed of the car.

Dedication:
The book is dedicated to my family, who always got together as a team whenever these annual Pinewood Derby tournaments came along.

This abridged book is one of two versions. This version is lighter and talks about general ideas, without going into any engineering analysis. This lighter version is more fun and appropriate for someone who is not interested in engineering equations. If you want the engineering analysis, then read 'Scout Pinewood Derby Cars and Real Cars, for Parents and Kids', which is for parents, kids, and engineers. The two versions are mostly the same except the heavier version has appendices with the analysis.

Copywrite © 2021 by Court Rossman
All rights reserved. No part of this book may be reproduced or transmitted in any form or by any means, electronic or mechanical, including photocopying, recording, or by any information storage and retrieval system, without permission in writing from the copyright owner.
Revision date: 4/07/2024

Who Should Read This Pinewood Derby Car Book for Kids

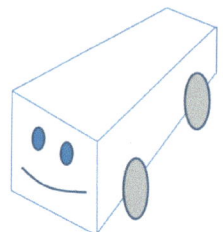

"Everyone can get something out of this book."

Here's who will get the most out of this book. Anyone, from kids to adults.

Elementary School student:

'Hey, this is a great opportunity to talk to my Mom or Dad about how to improve my car, and how it relates to people cars. And I like building stuff."

1. Techniques to make the car go faster.

Middle School student who is thinking about Engineering:

'Hey, I like these explanations, and want to learn more about them."

1. Techniques to make the car go faster
2. Hey, there are physical concepts that are working here, like conservation of Energy, except for all the drag from the monorail, the nail axle, and air.

High School student who is thinking about Engineering:

Read the version of this book with more analysis, called

Scout Pinewood Derby Cars and Real Cars: 'Dad, Sir Isaac Newton, and Me'

Format of book: yellow boxes are side comments about how pinewood derby can relate to real life designs, like people cars.

**Come on, give it a try.
Build a PD car, and ask questions.**

Table of Contents: Scout Pinewood Derby Cars, for Kids

Chapter 1: Who Thought Up the Pinewood Derby .. 6

Chapter 2: Cutting the Shape of the Wood Car 11

Chapter 3: Day of Pack Races, the Gathering 16

Chapter 4: Regional Races 22

Chapter 5: Max Speed Tuning,
and Drag Sources 24
1. Drive Straight Down Monorail — 36
2. Axle Friction: Graphite and Smooth Nail — 47
3. Avoid Air Drag — 52
4. Weight in Back Gives More Speed — 58
5. Lighter Wheels and Minor Stolen Energy — 62
6. Use Maximum Weight of 5 ounces — 67
7. High Bumper and Paint Job Minor Tweaks — 69
8. Dumb Luck and Foolishness — 70

Chapter 6: Fastest Track Shapes 72

Chapter 7: Odd Cars, Like Cars for Fun 75

Chapter 8: CO2 Power Cars 77

Chapter 9: No-rules 'Outlaw' Cars 80

Chapter 10: Summary Questions and Answers 90

Appendix A: Wheel Alignment with Shims or Nail Rotation — 93
Appendix B: Definition Energy and Force — 94
Appendix C: Racing Rules — 95

Pack Races

Regional Races

Wheels rubbing on monorail

1: Go straight: Align wheels

2: Less axle nail friction: Graphite powder

Outlaw cars with rockets and propellers

Gravity is Good (and Engines are Better)

What is a car like without an engine? Well, the car is called a Pinewood Derby car, or a go-kart. Gravity is good. It keeps you planted on earth. It lets you coast down a hill on your bicycle. It powers derby cars to accelerate down hill, and coast to the finish line. Unfortunately, gravity does not bring you back up hill. It takes engines and muscle to lift cars against gravity, which is why there are car engines on the highway and not derby cars, and why there are no jolly green giants lifting people cars up hills.

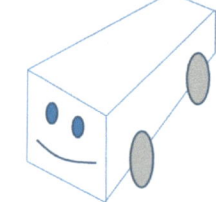

"Pinewood Derby is all about using gravity 'g', like apples falling from the apple tree."

Track for Gravity Powered Pinewood Derby Cars

Gravity powers the cars down the ramp, with no engine to get back up. Car accelerates from gravity, like a falling apple

Car coasts: For speed, the car needs to go straight, have low nail axle drag, and have low air drag.

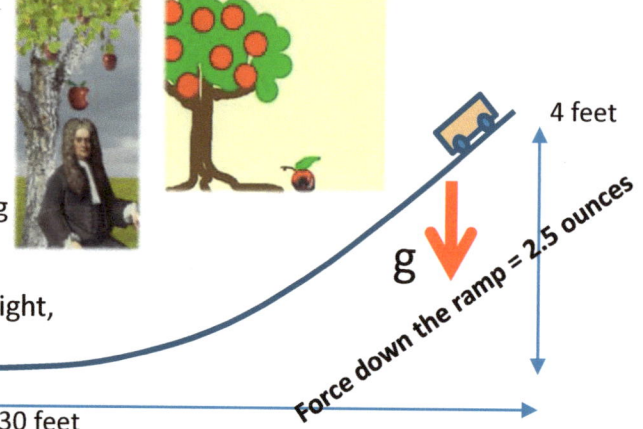

4 feet
g
Force down the ramp = 2.5 ounces
About 30 feet

Engines are better than gravity...
Real world comparison to engines

Real cars don't depend on gravity: Engines can go up hill (whew!...less hiking)

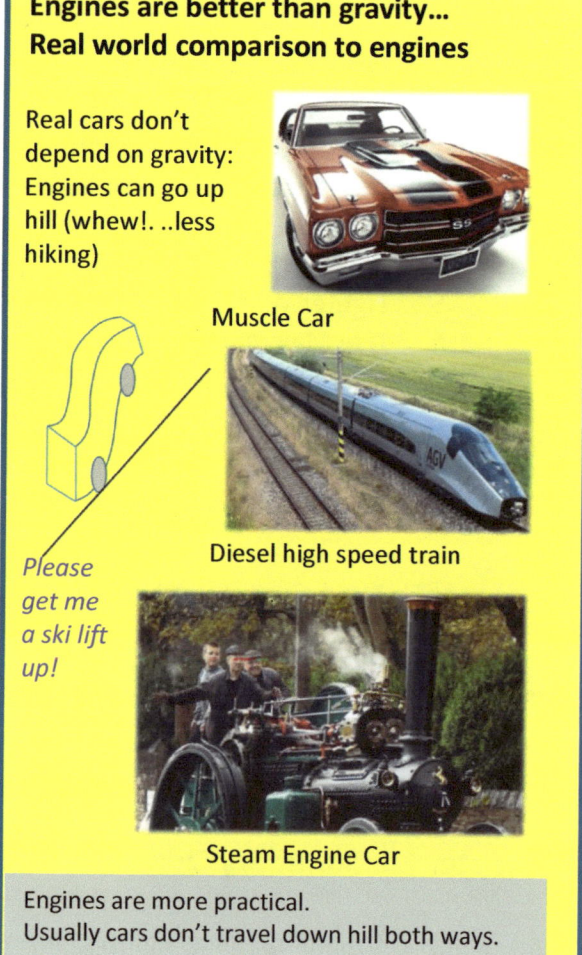

Muscle Car

Diesel high speed train

Please get me a ski lift up!

Steam Engine Car

Engines are more practical.
Usually cars don't travel down hill both ways.

What other sports use gravity and gravity power?

Sledding • Snowboarding / Skiing • Soapbox Derby • Trampoline • Luge • Ski jumping

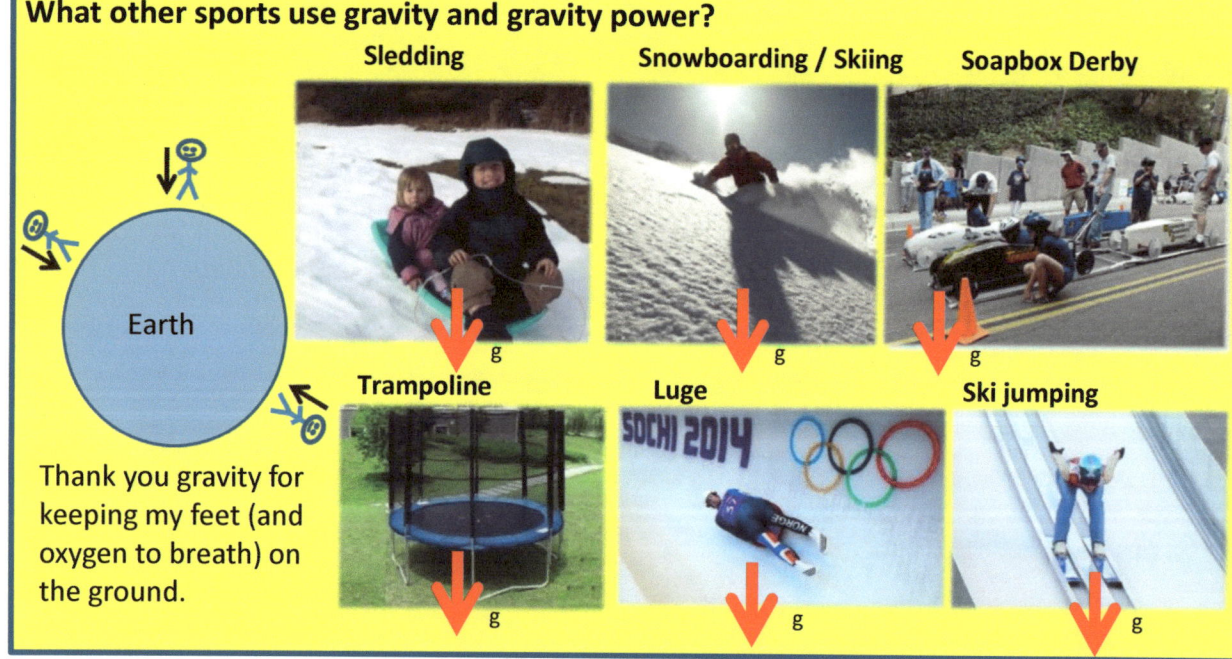

Earth

Thank you gravity for keeping my feet (and oxygen to breath) on the ground.

Gravity creates all sorts of fun, and keeps us on earth and alive.

Other Gravity Power, for Electricity!

We use gravity whether we want to or not. For these power plants, we want to. Water flows down hill and turns a turbine and a generator for electricity. The biggest gravity generator for heat is the Sun, which collapses all this gas until nuclear fusion happens.

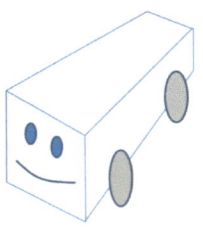
"Next time you scrape a knee, be grateful gravity is there to make you fall."

Electricity for wall sockets

Hydroelectric dam: 20% of world's electricity comes from water powered turbines, flowing downhill from gravity.

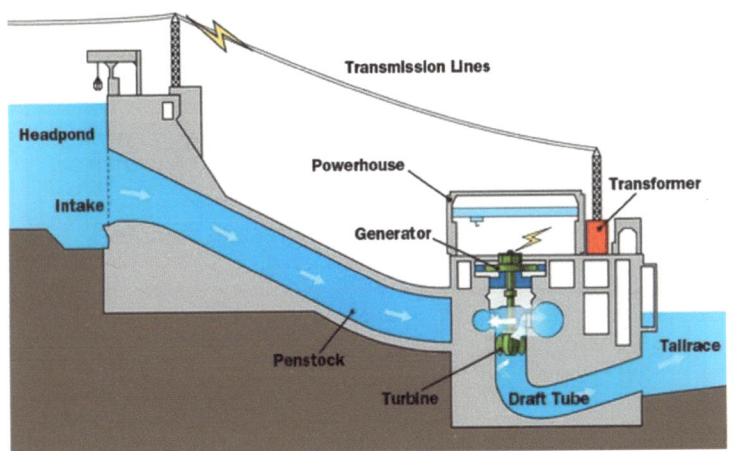

High pressure water from gravity, turbine, and generator

Mills to grind flower and cut trees

Water Mill: Gravity and water powered the world before 1900.

Solar System

Weight Power

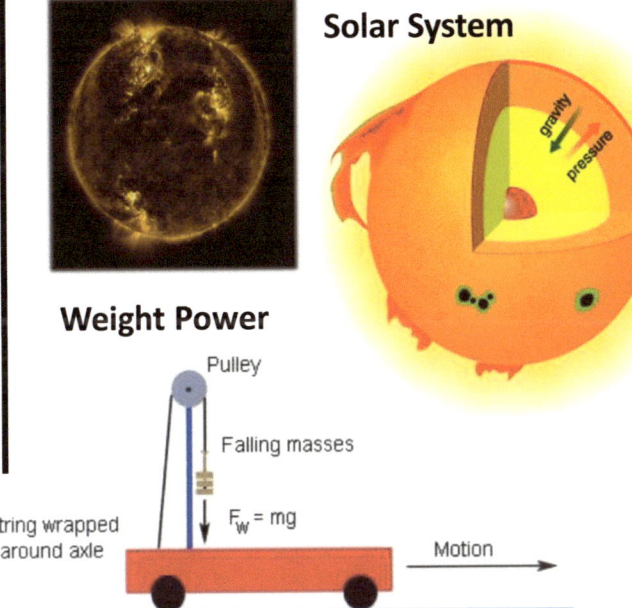

Other gravity cars are possible too!

The gravitational pull toward the Sun's center is balanced by pressure from heat caused by nuclear reactions.

Without gravity, the Sun's center would not get hot enough for nuclear reactions.

Also, the Sun would not have formed in the first place, just like planets would not have formed.

Gravity allows hydropower damns to work, from rivers flowing downstream. Gravity causes the nuclear reactions on the sun.

Chapter 1: Who Thought Up the Pinewood Derby?

Pinewood Derby caught on and has stayed very popular.

Don Murphy, Cub Master for Pack 280C in 1953, created the Pinewood Derby activity for his 10 year old son, to help parent-son bonding and to foster crafting skills with his pack. Scouts under 12 years old were not permitted to participate in the old soap box derby, so Don came up with this other event.

The Pinewood Derby activity was immediately accepted by the national BSA.

Another cub scout classic is small wood sailboats.

"Wood working never gets old, and pinewood is a perfect combination of easy to cut and strength."

Popular activities in the 1950s were also building go-karts and sleds. This pre-plastic time was before plastic toys became so common.

The popularity of Pinewood Derby was part of social changes. In the 1950s many people moved from the cities to the suburbs. Instead of going to plays and sporting events in cities, people needed to start hobbies, such as model plane, train and boat building.

Why has the wood working PD construction and competition survived the age of video games? Some hands-on craft activities are just fun. Pinewood derby also has friendly competition. PD construction is also an introduction to woodworking. Wood is a very practical construction material, and furniture and houses are still built with wood.

Other wood projects: Wood regatta race, and bird houses

Here are other games that have stayed popular for decades:

Popular games in the 1950s, which still have originality, staying power, and ingenuity:	Popular games in the 1980s, which still have originality, staying power, and ingenuity:	Popular games in the 2010s, which welcome online games:
• Model planes with wood • Electric trains • Play-doh • Gumby • Hula Hoop • Magic 8 Ball • Mr. Potato Head • Silly Putty • Tonka Truck • Monopoly / cribbage • Baseball	• Plastic model planes • Bicycling • Monopoly • Tennis, basketball	• Minecraft online play • Nintenko Switch for portable games • Fidget Spinner • Soccer

Pinewood Derby came from a time of more hand work, and it is still very popular even in competition with electronic games.

Vintage Pinewood Derby Cars

The kits had 'four wheels, four nails, and three blocks of wood'. The kits were modeled after racing cars in the 1940s.
In the 1980s, the kit changed from a rounded block to the current rectangular block, and the wheels were widened.

"Back then people really needed to smooth any flanges on the nails."

Original car styles for Pinewood Derby.

Car Racing in the 1940s

Rules 'back in the day' were about the same.
The older kits were actually more focused on the 1940s rally car.

Original Rules

The first rules in the 1950 prohibited lubrication on the nail axles. Most everything else is the same. There were even kits for sale. The kits had 'four wheels, four nails, and three blocks of wood'.
In 1977, the wooden struts were moved inward to the current positions.
In the 1980s, the kit changed from a rounded block to the current rectangular block, and the wheels were widened.

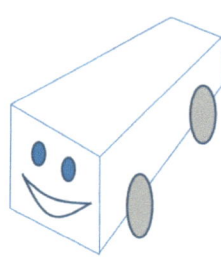

"Back then people really needed to smooth any flanges on the nails."

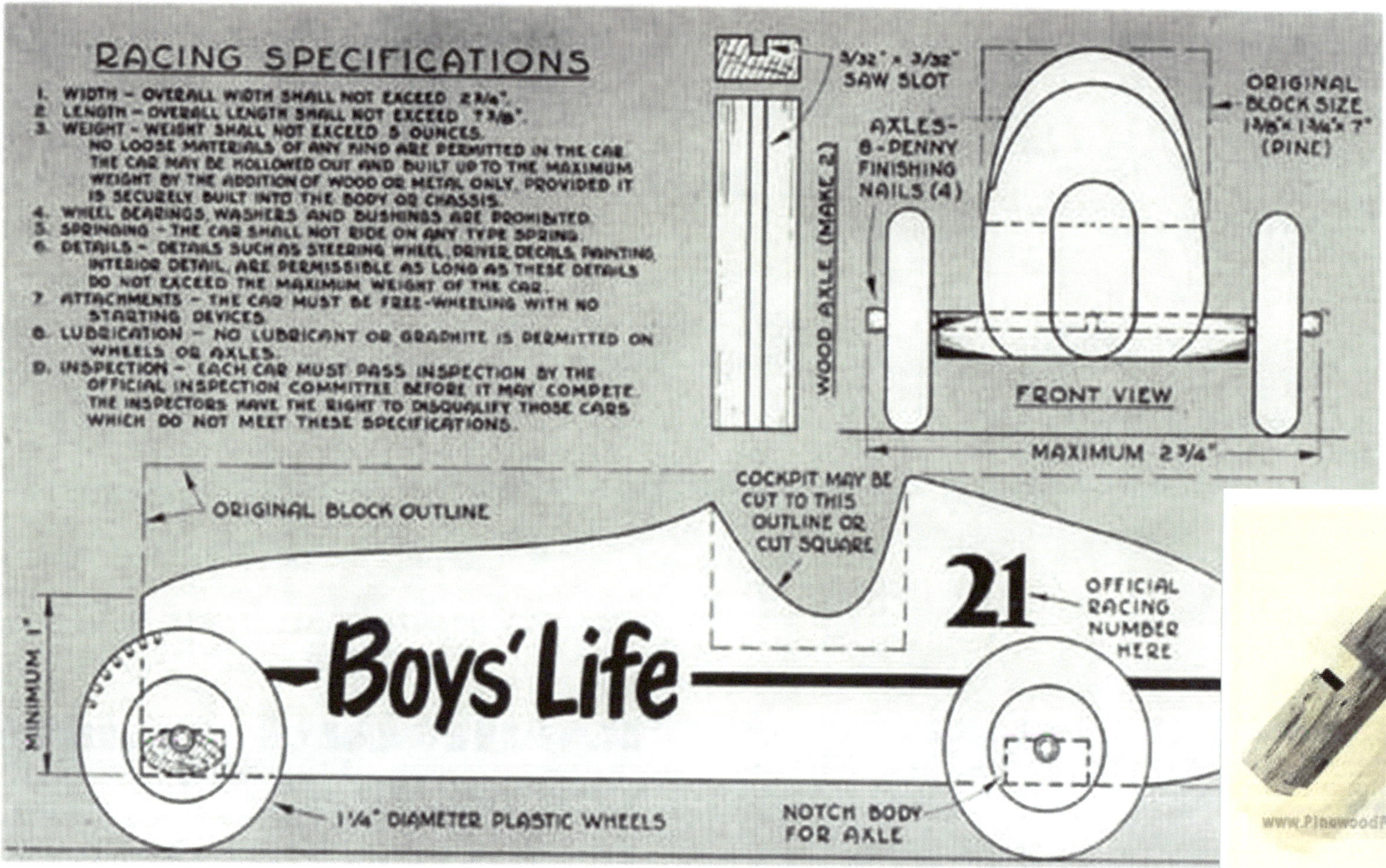

Original rules regarding length and weight, and no lubrication

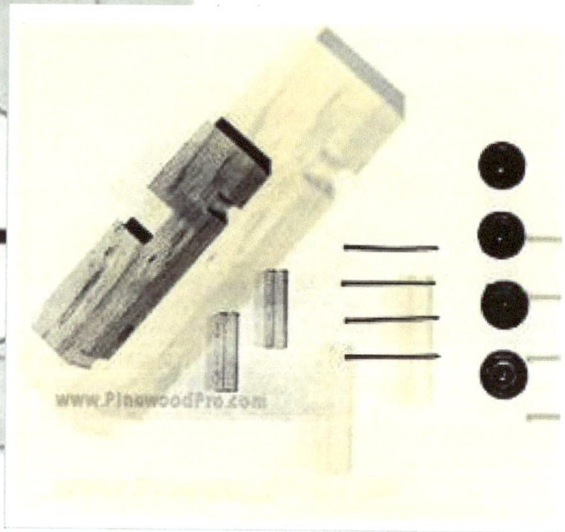

First kit in the 1950s

**Rules 'back in the day' were about the same.
The older kits were actually more focused on the 1940s rally car.**

Pinewood Derby's Long and Noble History

The first Pinewood Derby race was held on May 15, 1953 at Manhattan Beach

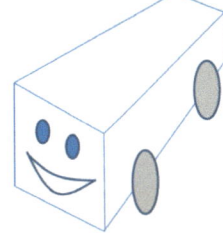

"These are the future race car and pickup truck owners."

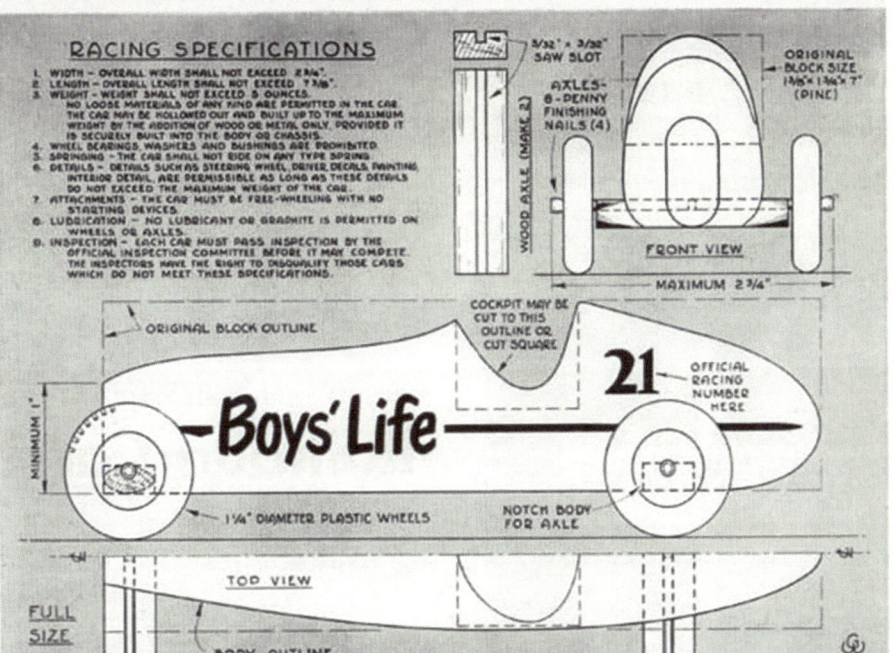

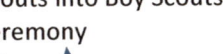

Pinewood Derby Poster to show cub scout achievements:
Presented at a graduation ceremony from Cub Scouts into Boy Scouts at the Blue and Gold ceremony

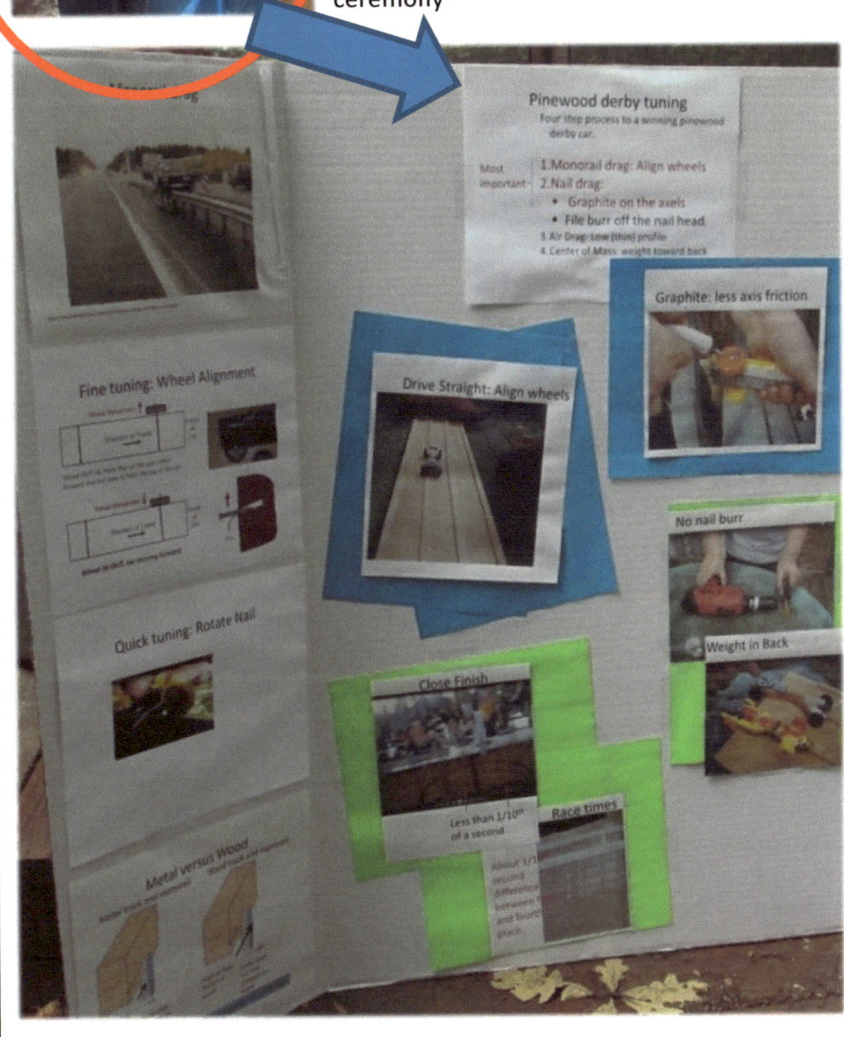

Pinewood Derby racing has been with scouting for over half a century. Other scout races, like toy sailboat racing, are newer.

Beyond Pinewood Derby: CO2 Rocket Cars and Soapbox Derby, Where People are the Drivers, No Engine

Cub Scouts get older, and what's next? CO2 derby, Soapbox derby in Boy Scouts. The Pinewood Derby is just the start of the fun: as the builders get advanced, they can upgrade their game and build and compete in a Soapbox derby.

Rocket High Pressure CO2 cars

Gas pressure CO2 car: The CO2 car can travel without gravity, as a rocket! You can race along the flat ground.

Rocket car: A burning rocket and a CO2 car have exhaust, so rockets and CO2 cars are the same idea.

Soapbox Derby (big gravity car with people)

Line up to start

20 feet of booster start from another scout

Go Kart cars: Start on top of a hill and use gravity and reach up to 35mph, for ages 10 through 20.

Building a Go Kart: Get some 2x6 inch boards, a threaded rod for axle, some lawnmower wheels, and one big turning bolt for the vertical axle pivot.

"Severe jealousy! I can't carry people, but I'm proud to train future drivers. ...and boy, do they need the training. In a soapbox derby without engines, the drivers weave all around from over steering with their front feet. Luckily, adults don't need to steer with their feet."

Go Kart races in Boy Scouts:
Warning: Steer, but don't over steer. You could turn too sharp and flip over.
Wear a helmet, and hopefully elbow/knee pads. You could screw in blocks to limit the turning of the front axle.

What's beyond Pinewood Derby? The future after Pinewood Derby gets real exciting. There are gravity cars with people in them!

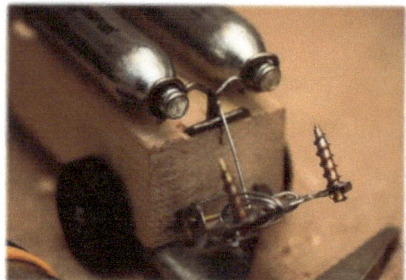

Chapter 2: Cutting the Shape of the Wood Car

*"Come up with your Grand design, and start cutting.
But don't adjust the axle grooves."*

You get handed a square wooden block. What are you going to do with it?

You still need a concept for the car. Some people just go for the as-is block, without any cutting, and just paint it. That works fine, as long as the car is tuned so the car goes straight and the nails are smooth.

You should grab a pencil and make a template of your design. Draw the cutout of your design in the block of wood. Let the saw follow the pencil markings. If you've never played with a hand saw or hammer before, now is the time to learn. Pine is a flexible soft wood that is easier to cut.

Do not play with the grooves for the nail axles. They are cut as close to 90 degrees as possible, to help the car go straight. Also, typical rules say that you can not use more spacing (wheel base) than what is provided.

The simplest design is to round the corners of the as-is block, or cut a wedge-shape profile into the block to lower air drag and bring the weight toward the back. Either way, even with a simple block, you need to add lead weights to get the weight of the car up to 5 ounces. With the maximum weight, you only get another ½ car length ahead, but any gain is something, and all those somethings add up.

Basic block

Custom Kit (a Jeep)

Here are the cars we built over a few years of cub scout Pinewood Derby tournaments, both 'street legal' and 'outlaw' cars. They are described in more detail throughout this book.
After a few years, you too will have a showcase of PD cars.
Each year the scout needs to build a new car. Older cars from previous year can not 'legally' compete.

Display case of many car experiments, with 6 years of cars and experiments, through many cub scout experiences.

Handy Hand Tools, Avoid Anything Electric

You'll need a few supplies.

It is helpful to know a person with a garage full of wood working tools. Still, if this is your first time making a Pinewood Derby car without a stockpile of PD parts, you'll probably need to make a pit stop at a local hobby or craft store to get some paints and lead weights.

"Grab some hand tools. Grab or go buy some Pinewood Derby gear (weights, paints, lubricant) at a local hobby shop."

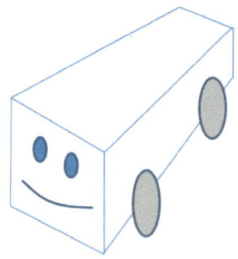

Pinewood Derby specific supplies:
- Paints
- Brush
- Various lead weights
- Scale for 5 ounce limit
- Light lubricant for the nail axles (graphite or Teflon, not oil)
- Glue to help keep nails in wood grooves (glue is poured in after car is built and tuned up).

General shop tools, to cut and shape wood without electricity:
- Saws
- Files
- Hammer for nail axle
- Sandpaper.
- Vise to bend nails

Optional electric tools (typically for the adult to use):

Drill and drill bits:
- Holes for lead weights
- Turning the nail to sand off the burrs.

Band saw:
- Quick way to hog out the shape of the car from the straight block of wood.

Vice: Slightly tap one nail to create a one degree bend on one nail axle.

Gather some regular wood working hand tools, like saws, files, sandpaper, and a hammer. A C-clamp or a vise is also useful.

Car Speed and Buy-at-the-Store Kits

When you buy a Pinewood Derby kit, you have a decision to make. You can go for your favorite truck, or you can go for speed.

Your truck design is taller and has more air drag. A lower profile car design has less air drag. The truck has more weight up front, so it has less gravitational energy on the ramp with a sloped start. Still, any car needs to roll straight or it will rub against the monorail. An as-is block car that rolls straight will easily beat an aerodynamic car that keeps turning and braking into the monorail.

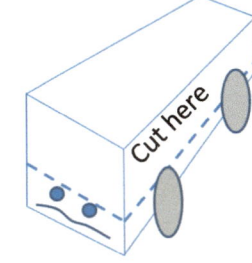

"I feel like someone is going to shave me to paper thin."

Basic Block:
Let your creativity and skill with a handsaw be your guide! Keep wood as a basic block, or cut it into something slim or creative.

Pinewood Derby Complete BSA Car Kit - The Lazer

Pre-cut Blocks for speed: Kits have given you a thin car for low-air-drag, with weight-in-back design.

Slower (taller and more air drag, weight too far forward) → **Faster** (slim less air drag, weight in back)

Basic Block of wood:
Woodland Scenics Pine Car Derby Car Kit, Basic

Kit of truck with more front area:
Revell Pinewood Derby Military Vehicle Racer, Officially Licensed Boy Scouts of America (BSA) Pre-Cut Shaped Wood Block Kit with Official Wheels and Axles

Kit of wedge with less front area:
Revell Pinewood Derby Wedge Basic Racer, Officially Licensed Boy Scouts of America (BSA) Pre-Cut Unfinished Wood Car Body Kit with Official Wheels and Axles

Slim and weighted cars:
Ultralite Pinewood Derby Car Body Only - Canopy #7 by Derby Dust

Slower:
- Taller so more air drag
- Weight up front instead of in back

Slower — Less air drag because slimmer — Faster

Slower — More weight in back — Faster

Faster:
- Slimmer so less air drag
- More weight in back

Slim is in, if speed is all you want—less air drag and can add more weight in back. Still, going straight and axle lubricant are more important than a slim car.

Car Assembly and Painting

True beauty lies in the eye of the beholder, so let's paint.

"You don't need to be a Picasso, just pour the paint on."

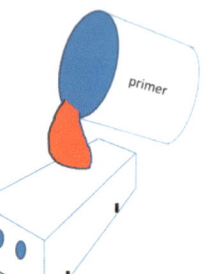

Starting block, for your imagination
Scouts get the official wood block and the rest is up to them.

Grab your derby box of materials
Materials collect over many years of competition.

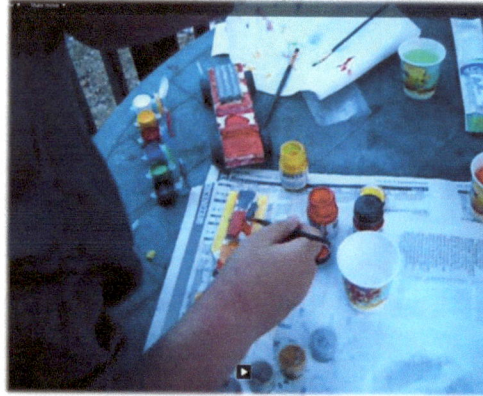

Dedicate a table for painting
A cheap plastic table outdoors, covered in old newspaper, can let messes happen without any hard feelings.

Fun facts: Robots now do the paint job for real cars

Painting a car in a factory: first a primer, then the color paint, then heating.

Car paint needs to survive years of weather (heat, cold, rocks, salt, ice) to protect the metal from rust.

The actual process of building the Pinewood Derby car:
- Get your gear in order: color paints, brushes, and newspapers, saws, hammers, lead weights, friendly encouraging attitude, creative streak, sand paper, wood files. You might need to go to an art store or a hobby shop.
- Remember, because straight wheels are much more important for pack races than air drag, you can have fun with the car and still be competitive. Make a Snoopy car, a banana car, or a skull car.
- Again, if parents want to help, they should build their own separate derby car beside their scout, so the scout does the work, especially as the scout gets older. That said, help your scout as much as they want, even on their own car, because this is a family bonding time.

Let the scout do what the scout wants to do

Why Pine?

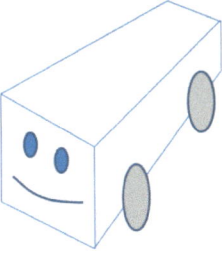

"Pine is a perfect combination of easy to cut and strength."

Pine is a soft wood that is easier to cut and sand than a hard wood. Pine is also flexible and less likely to crack. Pine is also common, available, and inexpensive. Pinewood used for PD cars is untreated.

Is Pine wood better than hard wood? What would the experience of cutting a hard wood look like?
Hardwood is brittle, so it breaks into slivers and bits.
Hardwood like oak is hard to cut with hand tools, so most likely a parent would do all the cutting and sanding.

Is Pine wood better than the softer balsa wood?
Balsa is lighter, but balsa wood is also easily broken. So after a few crashes the car is 'totaled'.

Here are the properties of Pinewood Softwood:

1. Pine is a coniferous wood that grows in many places in the norther hemisphere. There are over 100 types of pine tree species.

2. Pinewood is not brittle, and has durability, medium strength, and flexibility.

3. Pinewood does not require pre-drilling to attach screws and nails. Hardwoods like oak will crack when a fat nail is hammered into it, and pine will not.

4. Untreated pinewood has no resistance to rot, decay, and insects. So, if the wood is designed for outdoor use, pinewood is pressure treated to increase its resistance level. The pressure treatment forces preservation fluids into the wood. PD cars use untreated wood to keep the wood light and easy to cut.

Pinewood is strong and light but still softer than hardwoods.

Make paper, wood boards, wood particles for particles boards

Pressure treated: After preservation treatment, the wood is placed in a higher pressure container to force the treatment deeper into the wood.

Chapter 3: Day of Pack Races, the Gathering

Who's excited to race their derby creations? Everyone!
The scouts are always testing their cars down the track, before the final check in. Fast, slow, funny, accident prone, it is all interesting and an accomplishment for the scouts.

"Pack races are the climax of the cub scout season."

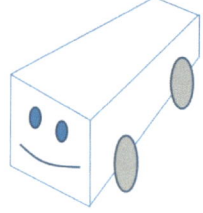

Finish gate, with light timers and drag ramps

Cub Scout Pack races, on a 3-lane wooden track

That's embarrassing. Maybe you don't want to be this guy celebrating, but you want to go to the regional tournament.

Faster cars

Low Profile Car (a custom kit), with weight in back

Most Original car

Trending movies

Funniest car

Music bells and whistles

Remember that the Pinewood Derby is about having fun, learning something, and sportsmanship: 'fun with a purpose'.

There should not be complete emphasis on the fastest car. Parents and kids would rather see the funniest car, and root for it. Beside fastest car trophies, there can be awards for most creative car, the funniest car, or the slowest car that still gets to the finish line. The scouts can look at the cars lined up on a table, and vote for one for each category.

Relax and goof off, and make a derby car at the same time.   Parents can make their own car side by side.

Final No-Excuses Check-in at the Races

"I sure hope I pass."

Going to check-in means handing over the car and declaring the end of the design, any design changes and tweaks.

There is no opportunity for unsportsmanlike modifications, like adding a few ounces of weight. The car is what it is, and you stand up and face the fun. You hand the car over, it passes (or not), and it goes into a box to be handed to the race operators. Maybe you'll be asked to bring the car back to the starting line after a race. You've invested your blood, sweat, and tears (or not) in it, and now the car performs as it will perform. Whether you win or lose, the car is out of your hands. No excuses! (unless you drop the car and bend an axle)

Check in table, with measuring tools

Your to-do list before Check in: (just before)

Reduce nail bearing drag: Douse the nail axles with lubricant just before check in. Usually re-applying lubricant is prohibited after check in.

Double check that the car rolls straight: If you've done some fun test runs on the troop's track, make sure your car didn't get banged up with misaligned wheels.

Check in for 'Street Legal' rules:

Car not too tall (5 inch). Too tall will hit finish gates. Think of trucks and getting under toll booths and bridges. A too-tall car can also tip over and damage cars on other tracks.

Not too long (7 inch). Too long and the car center of mass will be higher than the other cars, and the car will go faster at the bottom.

Enough ground clearance: Car bottom does not rest on monorail

Max weight rule: 5 ounce If more weight, then air drag and car design wouldn't matter so much. Designs should be close to max weight, without increasing front area and air drag, so that gravity force is much larger than air drag. Also, the forward energy of body of car is much more than rotational energy of wheels.

Distance between wheel axles must be between 4.0 and 4.5 inches: Longer axle separation will roll straighter, so there will be less rubbing and drag against monorail if car starts to turn.

- Rules are the rules: We tried longer wheel base at the troop level, but at the regional competition we either had to get dis-qualified or drill new holes, holding a portable electric drill as straight as we could.

Real car's annual inspection: Get used to it, you'll be having car inspections on people cars for years to come.
- Emissions
- Lights
- Tires
- Rust

Emissions

Headlights and blinkers

Traction on tires

Rust

The before-race scene is chaos, with pickup-up races. Make sure you've added light lubricant and followed the 'street legal' rules.

Both Pinewood Derby and Real Cars Have Rules

Different race activities and purpose means different rules. For any race competition—Pinewood Derby, or people cars like Nascar and Indy cars—we want comparable speeds to keep the race fun to watch.

"When I grow up, I want to be an old school Nascar!"

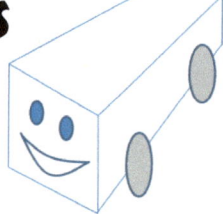

Pinewood Derby
Have rules for no engine to keep the competition even.

No engine, and 5 ounce weight

Rules:

Engine: None
- No engine or power on car
- Just use gravity

Weight: 5 ounce max
- Maximum weight of 5 ounces
- If heavier, air drag and rotational energy of wheels have less impact.

Safety:
- No loose parts that fly away
- No safety specifications, because no living passengers

Other:
- Maximum distance between axles. If longer, then less turning and wheel alignment less important.
- Max 7 inch car length, max 5 inch height.

See Appendix C for details on race rules

Regular Cars on the road
Have rules for safety and emissions.

Safety and Pollution

Rules:

Engine: Any
- Any engine that meets exhaust and pollution rules, but people like more acceleration which means bigger engine.

Weight:
- No rules, but lighter is better as long as safety and strength requirements are met.

Safety: Lots
- Things you'd expect on the road, like brakes and emergency brakes, mirrors, lights, and tire reliability
- Safety specifications to help people survive a car crash (roof strength, accordion collapse during crash, non-shatter laminated glass, bumpers) in the Federal Motor Vehicle Safety Standards.

Nascar (Stock Car)
Have rules for old school engines, safety and car weight.

V-8, Safety and Weight

Rules:

Engine: Old school, existing cars
- Large V-8 with an iron block, with 4 speed manual transmission, about 875 HP.
- No turbo-charge, to avoid turbo lag, too much power, and too much cost.
- Allow fuel injection.
- Starting to allow hybrid engines, to stay relevant with current stock cars.
- The starting stock car must have sold at least 500 same cars to general buyers on the road.

Weight:
- Minimum weight of 3450 pounds, without fuel, with no upper limit.

Other:
- Enforcement: Winning cars are torn down after the race to make sure the rules were followed (don't want too light, or too much horsepower).

Formula / Indy Cars
Have rules for safety and engine horsepower.

Engine size and horsepower, Traveling 230 mph

Rules:

Engine: Size and HP limits
- Engines must be 2.2 liter, producing between 550 and 700 horsepower.

Weight:
- Minimum weight of 1600 pounds, without fuel.
- If lighter, then the car has an unfair speed advantage because better horsepower to weight ratio.

Other:
- All cars are the same shape.

All races have rules.   People cars have rules that prioritize safety.   Nascar has rules that prioritize old style engines and power.

Good Bonding, Parents and Scouts

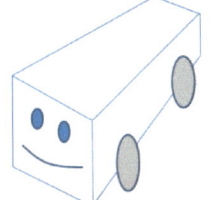
"I can't wait to race my friends!"

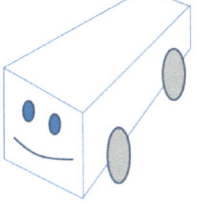
"I bring such pride and joy."

Bonding, learning skills, and friendly competition? This is the point, right? Good fun with friends, good fun with family, and learning about craftsmanship.
This might be the scout's first introduction to saws and sawing wood, or enamel paints and painting a car.   So be nice!

Sign Dad is taking the Pinewood Derby too seriously:
- Scout gets his first look at the car on his way to the registration table.
- Dad 'trash talks' the tiger scouts.
- 'Hey, there are trophies for adults too!'
- Dad writes a book on Pinewood Derby

Effort:
Design, cutting, painting, and testing

Awards for each year pack: Lion, Tiger, Wolf, Bear, Wolves, Webelos.

Races:
Excitement with the pack

Release cars by flipping the posts down, so the car start at the same time and height.

Reward:
A fun activity with awards

Main reward is hands on fun and learning, and good parent-son-daughter bonding time Also, there was a siblings/parents category for racing, so helpers can make their own car in parallel.

Any effort is good effort

Scouts and parents have a few weeks to cut the pinewood block into their creations.
A favorite car or creative shape is good. Fine tuning , like wheel alignment, happens afterwards.

Scoutmaster with race clipboard

Cars are lined up by the front bumper, so a longer car does not mean you cross the finish line first (although a longer car would allow the benefit of more weight farther back).

Scout gets award for first or second place in their age group

All age groups get recognition, even parents

Dad gets first place in parents category.

The scout is learning about cars, and learning that the parent is interested as well.

'Am I in Front?' and Old School Racing with No Timer

Most races are about finishing in front. Pinewood Derby uses gravity, and that means reducing drag. Real car racing uses engines, and that means lots of gas and a well tuned engine. For human sprinting, that means getting in shape.

"I'm a car length ahead!"

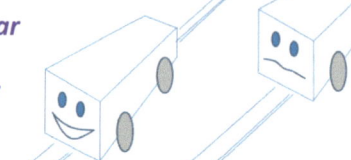

Head to head PD races

Pinewood Derby

Slim fast PD car on metal track

Some races are so close that a referee or a laser trigger is used.

However, either way, we don't need to know the total time, just which car crossed the finish line first or second.

Head to head races in other sports

Vintage car racing

Horse racing

Rowing

Nascar

Dog racing

Carnival racing

Formula / Indy Cars

Track sprinting

'Real Racing': Who's nose crosses the finish line first wins, regardless of rain or snow or temperature. The racers are all faced with the same track and weather conditions.
Qualifiers: These are typically timed, but the final race is not.

Pinewood Derby racing is very old school and traditional, where races are head to head and the fastest car wins.

Simplest Race Schedules Without Timer

Old school racing has no timer, just head to head racing between cars.
Watching the cars cross the finish line is unquestionable evidence of what car is faster. Even without a timer, some tracks have light switches that tell which car crossed first, second, and third.

"You need to sharpen that pencil and plan a race schedule that is fair and keeps all the scouts active, not waiting for 2 hours."

Double Elimination knock out without a timer

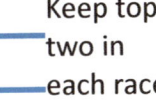 Can run each race three times, rotating lanes, so each car gets any smoothness advantage of each lane.

Keep top two in each race

Keep top two in each race

4-track Pinewood Derby race finish

- Only get accurate top 1 and 2 places, using a 3 lane track.
- Do not need a timer.
- If have a four lane race track, then can keep the top 3 in each set of cars.

Indy race close finish

Get some software and make it easy to plan the order of races.

Chapter 4: Regional Races, Where the Masters Meet

Did you get in the top 3 at the pack races? Now you are a master! You then compete in the big leagues against neighboring town packs a month or two later.

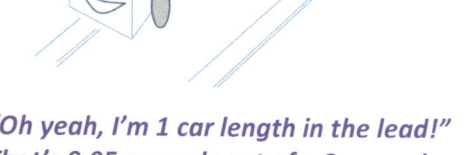

The regional races happen about a month or two after the pack races, where the top 3 or 5 cars and scouts from each pack get to go. This regional track was very nice. The track was metal, there are 4 cars side by side racing, and the cars are braked or slowed down after the finish line using drag on bottom and a bumper cushion. The big deal is getting to the regionals and having fun. The actual results at the regionals are not that important.

Regional competition, where top cars race from all packs.

"Oh yeah, I'm 1 car length in the lead!" That's 0.05 seconds out of a 3 second race, so we're all really close.

Real world: Close finishes in Nascar

Slowing down car after finish line, using bottom drag and cushion barrier, where all the energy goes into heat.

Close finish: Split second finish, only 1 car length separation

Less than 1/10th of a second, or about 1 or 2 car lengths

Race times: close top finish time of each car

About 1/10th second difference between first and fourth place.

Different Regionals have different setups.

If you are interested in winning, this regional race of top notch cars is very intolerant of imperfect tuning of cars. Everyone is at the top of their game, and have already finished near the top of their Pack. The car quality of the competition is great, and these are the best of the best. The cars must go perfectly straight to win, they must have low friction nail axles with axle grease, and they must have low air drag by being low profile.

Tiny tuning differences matter, as everyone approaches the gravity limit for fast times.

Smorgasbord of Entries to Regionals, the Best of the Best

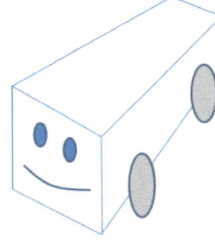
"We come in all shapes and sizes."

The Regional competition showcases lots of shapes and paint jobs.

We mostly have sleek low profile designs but some are just blocks and some have a toy driver. Even at regionals these are creative and funny cars, not just fast. Do you like fantasy cars, Nascar cars, Indy cars? You can find it all.

Partly these design variations show that design variations are a small factor in the speed. Rather, going straight and reducing drag against the monorail is king, and even a block car that goes straight will beat a sleek car that unfortunately turns into the monorail.

Not so shabby

Cars with upward exhaust, sails, Lego drivers

Cars with wedges like low-profile triangles, blocks like a brick, adorned for fun with windshields and eyeballs

Regional competition Pinewood Derby entries in Nashua, NH.

Car with side mufflers, and car with frame-only front end

Many good cars, and many good ideas

Chapter 5: Max Speed Tuning, and Drag Sources

The Pinewood Derby race is about using gravity to speed up and then coasting on the straight-away without slowing down due to any form of drag.

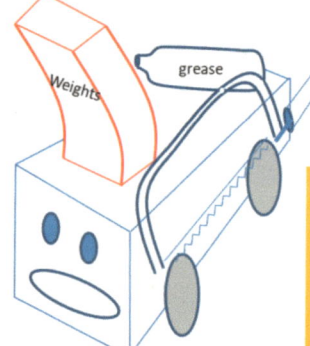

"Get your lead weights, wood saw, and nail axle grease!"

Race:
- The race lasts about 3.5 seconds, over a 35-foot track rolling down a 4 foot high ramp. The car's speed is 5 meters per second (m/s). If you scaled the small Pinewood Derby car up to people size cars, that would be 100 miles per hour (mph).
- After the car rolls down the ramp, the car is not gaining any more energy and the energy is getting sapped away from all forms of drag – from monorail rubbing, from nail axle drag, from air or wind resistance.
- Your goal is to remove all possible sources of drag. There is no fuel and engine, so you can not add any energy. All you can do is remove drag.

Use your knowledge for speed, not evil. Here are 'tricks' to get speed out of that lump of wood, and reduce drag. Make the most of gravity and don't waste energy in any form of drag and heat.

Six step process to build a winning Pinewood Derby car:

Reduce drag:
1. **Monorail rubbing and drag: Align wheels**
    - Don't crash and bend wheels before races
2. **Nail axle drag:**
    - Graphite on the axles
        - Sand grooves in nail to store the Graphite
    - File flange off the inside nail head.

(Most important improvements)

3. **Air Drag or wind resistance:**
    - Low (thin) profile
    - Wing-like shape

Increase energy slightly:
4. **Center of mass in back and higher for more gravitational energy:**
    - Weight toward back
    - Remove wood from middle so more weight in back
5. **Use all the 5 ounces allowed to plow through air drag**
6. **Lighten wheels to reduce spin energy, and increase forward energy.**

List of tricks

These tricks can be shared by anybody.
For example, here's a plug for another book that first enlightened me to a higher level of Pinewood Derby design: 'Pinewood Derby Speed Secrets', by David Meade.

Boy Scouts also published 'Pinewood Derby Master Mechanic Kit', by Paul Beck, which describes the engineering tips.

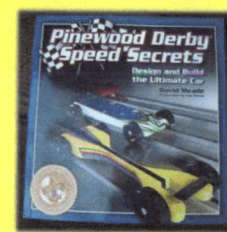

Go straight, go straight, go straight... and add lots of axle lubricant before handing the car over at check in.

Pinewood Derby Above-Board Tricks

Here are the main 'tricks' to get a fast car, or to stop your car from losing energy, mostly to avoid drag. Half this book is spent talking about these tricks.

"Learn it, Live it, Love it."

1: Go straight and don't rub against monorail:
- Align wheels so car coasts straight

Turning and rubbing on monorail

Rotate Nail: Quick alignment if nail is slightly bent

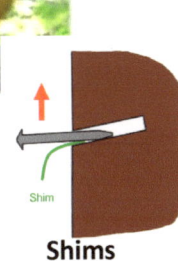
Shims

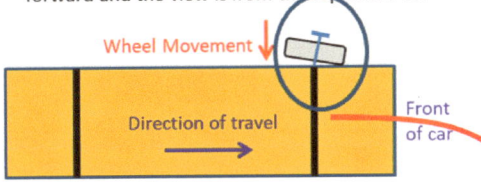
Wheel OUT-IN, Here the car has just rolled forward and the view is from the top of the car

Wheel IN-OUT, car moving forward
More accurate alignment:

2: Less nail axle friction:
- Graphite powder
- No nail burr

Graphite lubricant

Smooth nail axle, no burrs

3: Less air drag:
- Low profile car

Low profile and back weight

4: Keep weight in back
- Center of mass is higher on starting ramp, for more gravitational energy

Back weight

5: Use full maximum weight of 5 ounces
- Plow through air drag easier, with same car area

"I'm going to ram through that air with all my weight."

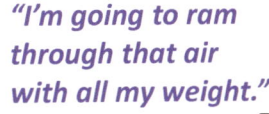

6: Lighten wheels

Close Finish

Less than 1/10th of a second, or about 1 or 2 car length separation

Every trick matters to keep those small fractions of speed.

http://www.pinewoodprofessor.com/friction.html
http://en.wikibooks.org/wiki/How_To_Build_a_Pinewood_Derby_Car/Assembly

List of tricks

If You Conserve Energy, No Drag, You Win

We know the starting energy. That is the energy from the height, or from gravity. There is no other energy because there is no engine. We want to keep that energy, so the speed at the bottom of the ramp has an energy equal to the starting energy.

Starting energy up top from gravity: The energy at the bottom can not be more than this.

Finishing energy at bottom in speed: The energy at the bottom can not be more than the starting energy.

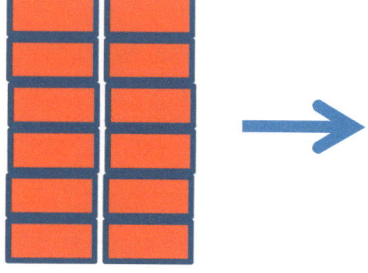

Potential Energy from gravity at top
Potential energy
= mass*gravity*height
= m*g*h

Kinetic Energy (motion) at bottom if no friction
Kinetic energy
= $(1/2)$*mass*speed^2

Energy lost to friction (monorail rubbing and drag, air drag, axle drag)

Diving, with some air drag

Diving: Falling divers pick up speed as they fall from the board.
- Friction is air drag.
- High board divers want air drag to limit the speed of their fall to just 100 mph.
- Air drag and speed are hardly the point. Divers just don't want to belly flop into the water.

Sky Diving, with lots of air drag

Air drag force: Air drag causes skydivers to reach a maximum speed and can't go faster (terminal velocity). The parachute has a huge air drag. Air drag also steals the gravitational energy. Again, people don't want to go faster, so air drag saves lives here.

Throwing a ball up and falling down (football, basketball), with air drag

Thrown Balls: Forward speed of ball is only steadily reduced due to air drag.
The football slows down its vertical speed as it gets to the top or apex of the throw. Then the ball accelerates back down.
- Friction is air drag.
- Footballs have a streamlined shape, with a low air drag factor.
- Baseballs have a bulky shape, with a high air drag factor.

Throwing and swinging a bowling ball, use muscle and let gravity help as well, with spin.

Bowling Balls: Bowlers swing the ball backwards and then force it down using muscle and gravity, pivoting on the arm. The ball would still have some speed due to the height it started, although using muscles helps too.
- Friction is air drag.

The goal of PD is to remove all loss mechanisms and preserve the gravitation energy down the ramp.

Gravity Energy at Top of Ramp is Wasted as Heat at End

Where does all the energy go? Say you bicycle up a hill and coast down to the same spot, standing and looking back up the hill. What happened to all that muscle and effort? Heat! You just heated up the earth a little, through air drag, bearing friction, tire compression, and body heat.

"Hey, is it hot in here? My wheels and face feel hot!

My wheels are hot from monorail drag, and my face is hot from air drag."

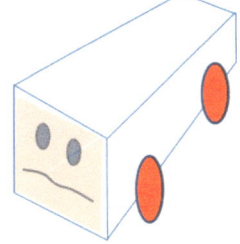

Energy from gravity at top goes to movement at bottom until car hits finish line, stops and generates heat.

And now for some great energy concepts!

Heat the monorail
- Waste energy, from friction with the wheels rubbing against the monorail (>10%)

Heat the plastic wheels
- Waste energy, from the nail friction (10%), and friction with the monorail (>10%): 20%

Heat the air
- Waste energy from air drag (8%)

Rotational energy in spinning wheels:
- Before all the energy is lost after hitting pads after the finish line, some energy does not go into forward motion: energy goes into rotation of wheels and wastes energy, instead of helping forward motion: 4%

Heat car and bumper after finish line (100%):
- At the end, past the finish line, the car hits a bumper and stops. All the energy goes into heat to stop the car, just like a car crash or braking a people car with hot brake pads.
- Pinewood Derby cars have lost all this energy or mojo when they hit the bumper. They were just a little furnace that helped to heat the room. The person that raised the car to the top of the ramp is an even bigger furnace.

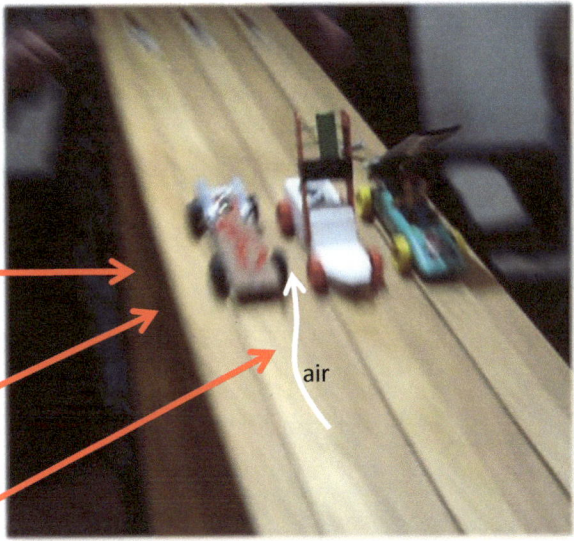

Who wins and loses depends on who has less drag from the monorail drag, the nail axles, and air.

Real world comparison to cars

Real people cars are bigger and waste much more heat than PD cars:
- Air drag = Heat of air and car surface
- Braking with pads = Heat of brake pads
- Road resistance or wheel compression = Heat of tire
- Losses in transmission = Heat oil
- Wheel rotation: does not matter, because energy helps keep car moving when already up to speed.

Standard sedan with engine has even more losses than a gravity powered PD car.

Like Pinewood Derby cars, motion of People cars with engines all goes to heat too. When you park your car in the driveway at the end of the day, where did all the gasoline energy go? Heating the engine, the air, your tires, your brake pads.

To slow down, an electric car with re-generative braking can convert some rolling energy into charging the battery, but not much.

The energy unleashed by gravity is, in the end, all wasted as heat. People cars also give all energy to heat after they stop.

Fast and Slow Car Ideas, by Design

Buck the trend: Go slow with class, using slow car ideas. Let's ignore poor tuning ideas like going crooked with more monorail drag, and ignore forgetting axle grease, which can slow even a 'fast' design car down to zero. Instead, let's look at how to make a naturally slow Pinewood Derby car, where the energy of gravity is less, or there is too much energy going to spin up the wheels, or air drag is huge.

"I'm in no rush. Let's just pop out a juice box, or the bubbly."

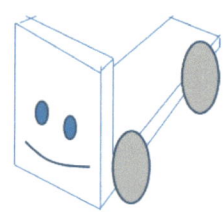

Fast car ideas, using the physics to win races

Weight in back

Fast car kit with all the tricks.

Low profile for low air drag

Assume wheels are aligned and greased.

Slow car ideas, to really demonstrate all the physics, creativity, and fun

Big Wheels

Fat, heavy wheels (too much energy in spin instead of forward motion)

Creepy Wheels

Lossy wheels that flex and heat (too cushy)

Front Heavy
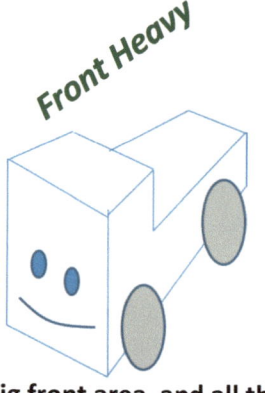

Big front area, and all the weight in the front (large air drag, and weight in wrong spot)

Butt Dragger

Remove rear wheels, and put cloth underneath as skids (sliding friction with top of monorail).

Air Wall and Under Weight
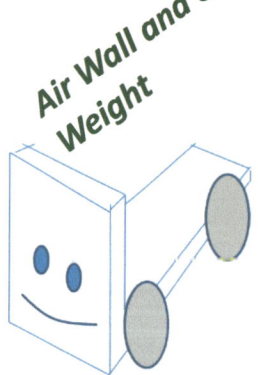

Big front area, and all the weight in the front, and below max weight, say 2 ounces (large air drag, less gravity force)

Double Air Wall
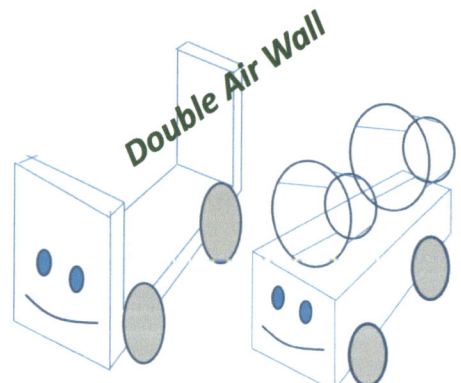

Double the air drag with two flat surfaces, or two Dixie cups

There is nothing wrong with proving cars can be slow (deliberately) as well as fast. People might root for it, if you add a pin wheel or add huge wheels made from CDs.

Loss Mechanisms for Gravity Power

There are many hodge-podge ways to lose energy, unfortunately. We need to tackle each one, mostly monorail drag (go straight), axle drag (axle grease), and air drag (low profile).

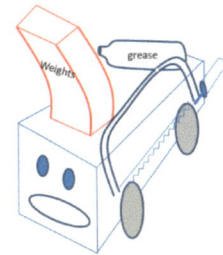

"Huge energy stolen through drag …you can keep only 30% of the gravitational energy, if monorail drag, axle nail drag, and air drag all kick in."

Gravity powered Pinewood Derby cars lose more than 30% of their energy from drag and heat, even with a well tuned design

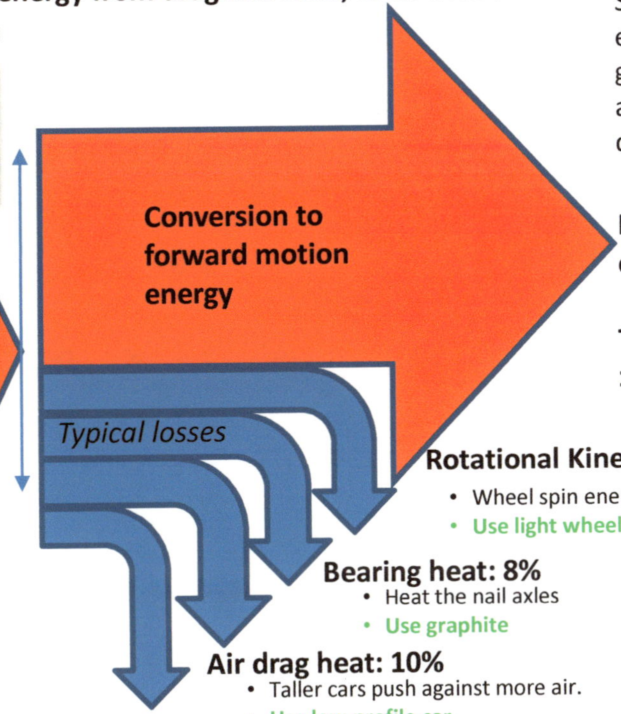

Gravitation Potential energy (1.7 Joules) → **Conversion to forward motion energy** → **Forward Kinetic energy: <68%**

Still quite a lot of energy kept if car goes straight and avoids mono-rail drag.

Total losses: >32%

Typical losses

- **Rotational Kinetic energy: 4%**
  - Wheel spin energy
  - Use light wheels
- **Bearing heat: 8%**
  - Heat the nail axles
  - Use graphite
- **Air drag heat: 10%**
  - Taller cars push against more air.
  - Use low profile car
- **Monorail heat: >10%**
  - Rubbing against monorail.
  - Go straight, Align wheels

The energy at the bottom equals the energy at the top, minus the air drag, the bearing friction, and other friction losses like monorail drag.

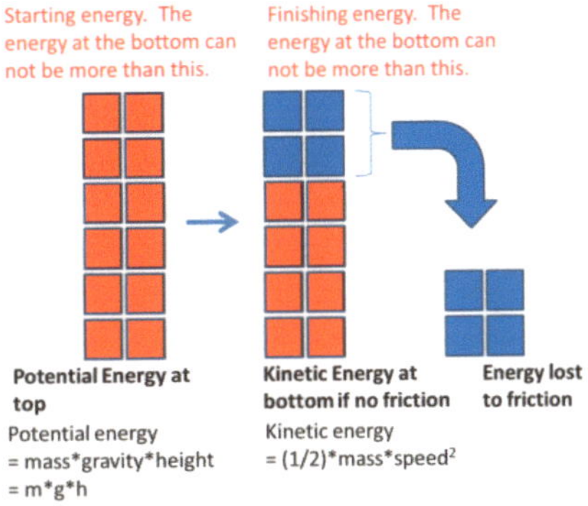

Starting energy. The energy at the bottom can not be more than this.
Finishing energy. The energy at the bottom can not be more than this.

Potential Energy at top
Potential energy = mass*gravity*height = $m*g*h$

Kinetic Energy at bottom if no friction
Kinetic energy = $(1/2)*mass*speed^2$

Energy lost to friction

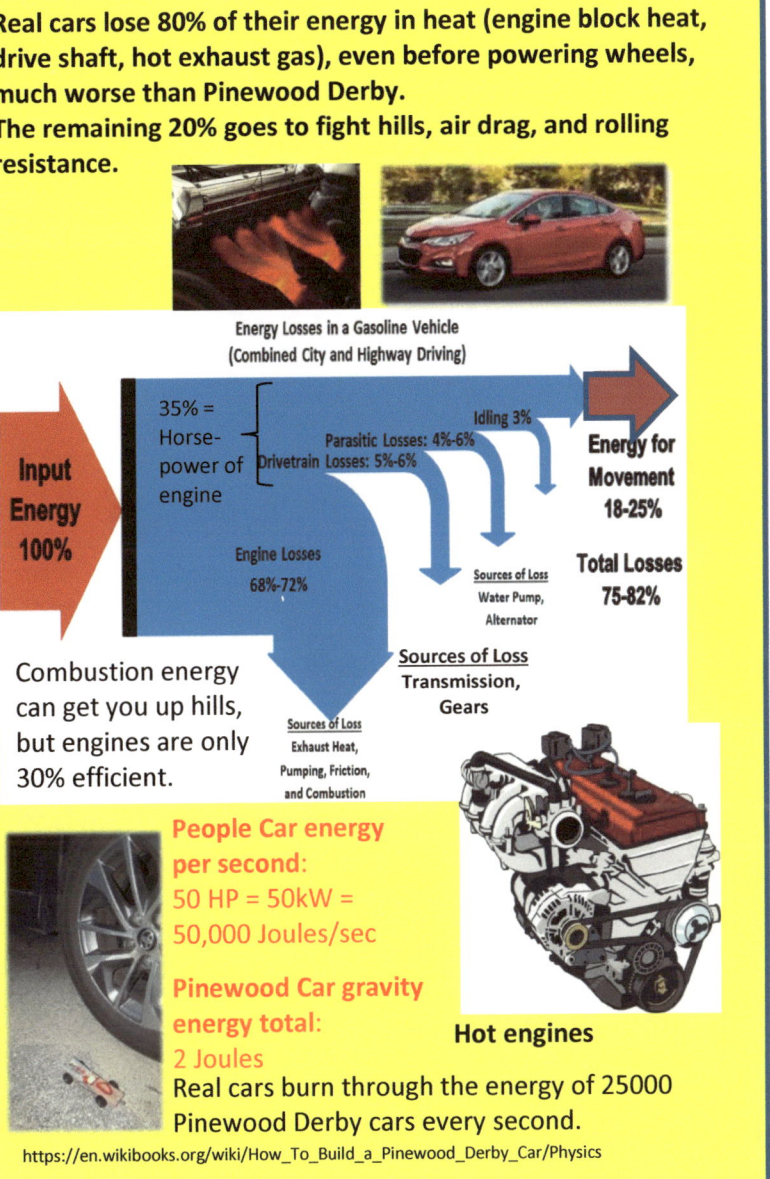

Real cars lose 80% of their energy in heat (engine block heat, drive shaft, hot exhaust gas), even before powering wheels, much worse than Pinewood Derby.
The remaining 20% goes to fight hills, air drag, and rolling resistance.

Combustion energy can get you up hills, but engines are only 30% efficient.

People Car energy per second:
50 HP = 50kW = 50,000 Joules/sec

Pinewood Car gravity energy total:
2 Joules
Real cars burn through the energy of 25000 Pinewood Derby cars every second.

https://en.wikibooks.org/wiki/How_To_Build_a_Pinewood_Derby_Car/Physics

All the loss contributions steal speed and steal up to 1 second of race time – many car lengths – from your car.

Typical and Improved Time Lost on Track

Here is a chart of typical amounts a Pinewood Derby car is slowed down due to each loss mechanism, monorail drag, axle drag, and wind resistance.

"Align those wheels, first and foremost! Then grease the nails."

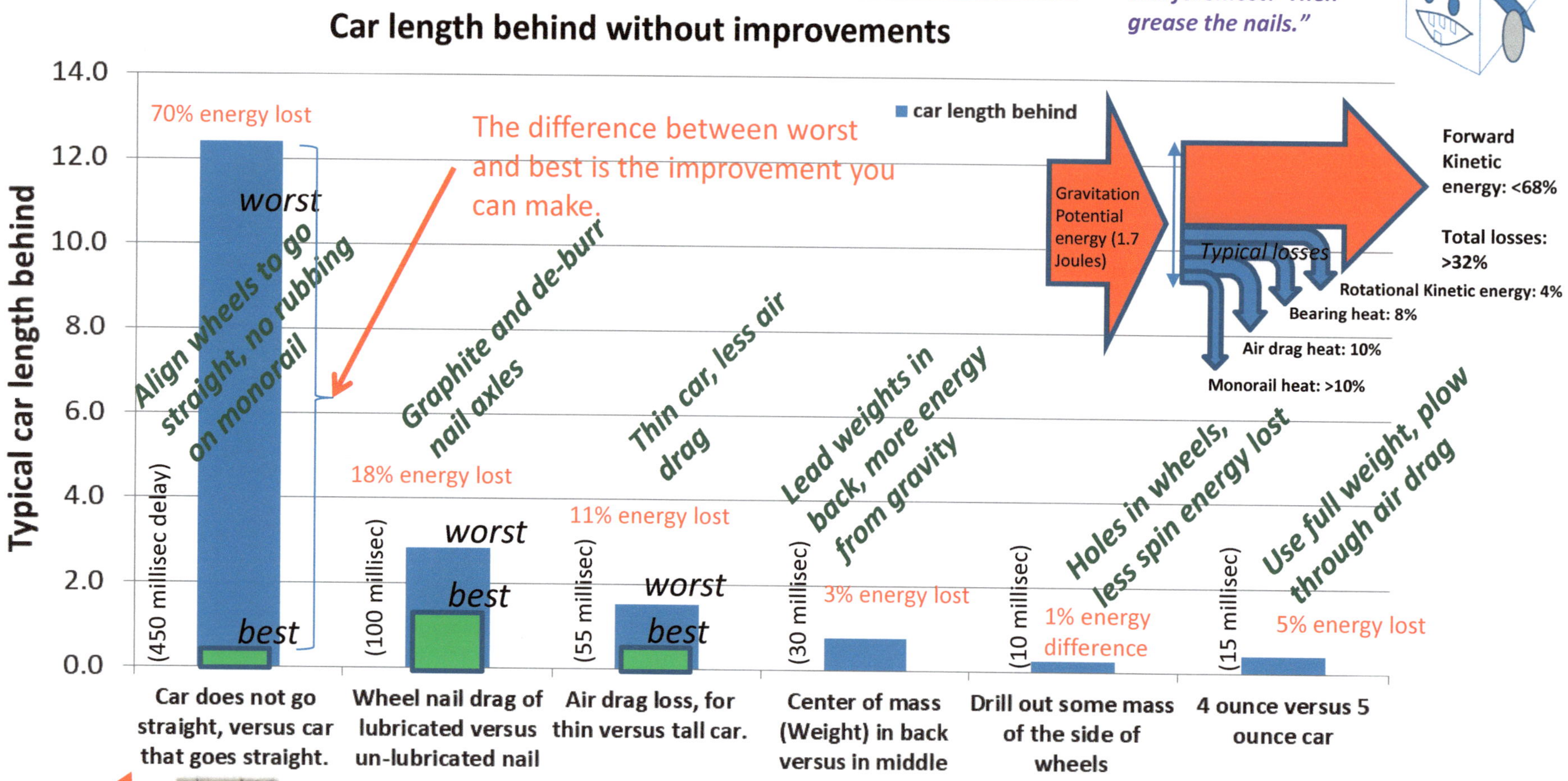

Car length behind without improvements

- Bar 1 (Car does not go straight, versus car that goes straight): worst 70% energy lost, best (450 millisec delay). *Align wheels to go straight, no rubbing on monorail*
- Bar 2 (Wheel nail drag of lubricated versus un-lubricated nail): worst 18% energy lost, best (100 millisec). *Graphite and de-burr nail axles*
- Bar 3 (Air drag loss, for thin versus tall car): worst 11% energy lost, best (55 millisec). *Thin car, less air drag*
- Bar 4 (Center of mass (Weight) in back versus in middle): 3% energy lost (30 millisec). *Lead weights in back, more energy from gravity*
- Bar 5 (Drill out some mass of the side of wheels): 1% energy difference (10 millisec). *Holes in wheels, less spin energy lost*
- Bar 6 (4 ounce versus 5 ounce car): 5% energy lost (15 millisec). *Use full weight, plow through air drag*

The difference between worst and best is the improvement you can make.

Energy flow:
Gravitation Potential energy (1.7 Joules) → Forward Kinetic energy: <68%
Typical losses:
- Rotational Kinetic energy: 4%
- Bearing heat: 8%
- Air drag heat: 10%
- Monorail heat: >10%

Total losses: >32%

Straight coasting  | **Axle grease**  | **Low profile**

Don't forget Pure Dumb Luck
- Aligning the car straight along the track at beginning (first few feet will rub against monorail)
- Getting on quickest lane, which is why cars are switched between lanes between races. Some of the lanes have rougher seams.
- Not dropping car before the race (mis-align the wheels)
- Make sure your scout doesn't try to bang up the car before the races, during practice runs, which kicks out the wheel alignment and lubrication.

Monorail Drag: You can make a huge improvement here, by reducing turning and drag on monorail. Worst case monorail drag can actually be much worse, if track is wood, and wheels are completely not aligned and car wants to do a severe turn.

Inside Wheel Drag against side of car wooden block: A bent nail also pushes the wheel against the block or nail head, adding to friction loss.

Go straight, go straight, go straight, and add lots of axle grease before handing the car over.

Monorail Drag is Rough Sliding Like Brakes, Not Rolling

Monorail Drag: Most of the friction and drag that slows down the car is due to the plastic wheel pushing against the wood or metal monorail, if the car doesn't roll straight and keeps rubbing.

Monorail drag is just like applying a car brake. In a car, brake pads push against the disks inside the wheel. In a PD car, the wheel is pushing itself against the side of the monorail.

"I want to go straight, and not feel the burn of my wheels against the monorail!"

Rollerblades also use rubbing to slow the rider down.

Soap box cars push some stick or rubber against the ground, or against the wheel like it is a disk brake.

Plastic wheels against wood:
More drag
Wood track and monorail

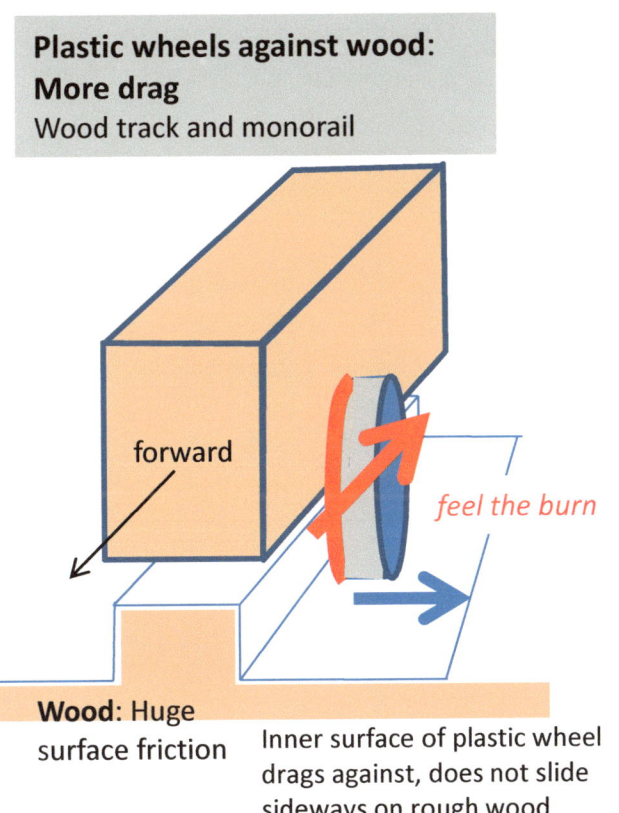

forward

feel the burn

Wood: Huge surface friction

Inner surface of plastic wheel drags against, does not slide sideways on rough wood

Car disk brakes.
Those brake pads and rotors get very hot. All that forward motion energy goes into the concentrated heat at the rotors. Some rotors warp from the heat if they get too thin.
The brake pads are made from organic composites (glass fibers, rubber, carbon and resin). A decade ago, brake pads were made from asbestos, but that material creates carcinogenic dust.

Rubbing against a wheel is used in car brakes, by rubbing against the disks. The disks heat up and air flow needs to cool them down.

Reduce Losses in PD or People Cars

"PD cars have some of the same design issues for loss that people cars have."

PD cars and People Cars each have their own issues, to get the most from the car.

PD cars need to go straight to avoid monorail drag, and a few other tricks. But PD cars are designed to go straight.

People cars have a lot more to build and to optimize. The engine needs to be efficient. The wheel bearings need to have low resistance.

Pinewood Derby cars, tuning

Issue	Fix
Rubbing wheels against monorail	Align the nails so the car coasts perfectly straight. A PD car can not turn with a steering wheel, but it can go very straight.
Going up hill or just along flat ground	There is no fix. The PD car is more like skiing downhill, or like a soap box derby.
Nail drag	Apply a light grease or powder lubricant to the nail.
Air drag	Use low profile car.
Wheel friction and losses against ramp	The wheels are hard plastic, do not deform, and don't have any losses from compression during the race.
Center of mass in the back	Place lead weights in the back above the back wheels, to increase the gravitational energy.

People Cars

Issue	Fix
Dragging wheels: Brake pads rubbing against the brake rotor	Replace the brake pads, unstick the pistons pushing on the pads. Re-spin the rotor so it spins flat.
Basic motion: Driving anywhere, up, down, or just a flat roads	People cars have engines.
Bearings: Lubricating the wheel bearing with grease	People cars use strong roller bearings, and there is heavy grease in the bearings.
Air drag: Pick the low drag shape. Unfortunately, real cars like pickup trucks have lots of air drag	Some cars can be low profile, like 2 seater sports cars, but most cars need headroom, like sedans, SUVs, and pickup trucks.
Wheel compression loses, which provide traction but do heat the tire.	Cars want some wheel compression to get grip, so some heating of the tire is inevitable.
Center of Mass: The center of mass is typically where the engine is. Some cars have their center of mass in the front with a front engine. Some sporty cars have their engine in the middle for better turning.	The engine is typically in front because the engine will provide a safety cushion in a frontal accident, using an accordion compression of the front. The tires which are under the heavy part of the car, like the engine, typically have the most grip.

Pinewood Derby cars need some tender care to go straight and reduce monorail drag losses.

Compare Drag of Pinewood Derby Cars and People Cars

Here are some differences from a PD car to a People car, besides having no engine, no driver, and can't turn!
- Pinewood Derby cars can be a scaled-down People car shape, but can be lower profile than real cars.
- Pinewood Derby axles probably have more friction than a real car bearing, and Pinewood Derby axles also need to be re-lubricated with graphite every 20 races or so. Real car bearings, with ball bearings and grease, can last for more than 100 thousand miles before adding more grease for the bearings.
- Pinewood Derby cars have hard plastic wheels that drag against the side of monorail, while real car rubber wheels flex and heat, to provide more traction and to make the ride more smooth for the passengers. Comfort versus efficiency. Comfort versus efficiency? You decide.

"These are real design issues for real cars."

Height of car and air drag:

Real cars need to carry people and are taller with more frontal area, so they have more air drag.

A scaled up Pinewood Derby car has about the same frontal area as a People car.

A sports car, like a Corvette, can be super low profile with the driver practically lying backwards, but not many people buy that.

Pinewood Derby car

People car

Wheel bearing friction:

A greased nail axle could not last long around corners and fast speeds. Imagine long chariot races where the wheel falls off.

For regular heavy cars (or skateboards), compared to greased nail axles, bearings should have much less bearing friction for heavier weights of the car.

Real cars have custom bearings to handle turns: the bearing has a taper to the roller bearing to withstand the side forces during a turn. That is, the bearing has a cone shape.

Cylindrical bearings to handle turns

Ball and cylindrical bearings roll. They don't slip like axle nails.

Fixed, non-rotating axle which bearings roll around

Spinning wheel

Wheel road friction:

Rubber tires compress, which heats the rubber and takes away energy. Plastic derby tires do not compress, so they should have less drag.

- Dirt bikes have fat tires, because tires need to grab mud, rocks, for traction, but fat tires have more compression and are more tiring to ride on.
- Road bicycles have narrower tires with higher pressure, so there is less compression, less grip, and less loss. Narrower tires are easier to push but less comfy ...most people want a cushy seat.

Cushy rubber car tires with more drag

Rubber tires compress and distort, losing energy, but the ride is more comfortable, and the grip is better in all weather and rough roads. Rubber tires especially distort during turns, which slows the car down (as they drag around their vertical axis).

Height, wheel friction, tire compression are all part of real car design.

Fast and Slow designs

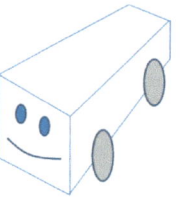

Every sport uses some concepts of fast and slow.
Boats have a huge amount of water drag, unless the boat can go so fast that it can skim across the surface like a water skier.
Bowling balls have a huge energy converted to spin, not just to forward motion.

"Pine is a perfect combination of easy to cut and strength."

Water drag: Boats have much more water drag than cars have air drag. Water is 1000 times more dense than air, so it takes more force to plow through it. To go faster, the hull should be skinnier and ideally plane or skim across the surface, using hydroplanes on underwater wings or skipping across the surface. A boat is much less efficient in terms of miles per gallon compared to cars.
For a PD car in air, a low profile pushes less air away and has less air drag.

Heavy displacement long-keel hull: lots of water drag

Hydroplaning: reduced water drag

Planing hulls: reduced water drag

Energy lost to spin: Things that need to roll have part of their energy in the spin, not just in the forward motion. To go faster, this spin energy should reduced.
For a PD car, lighter wheels, much less than the total weight of the car, reduces energy lost to wheel rotation.

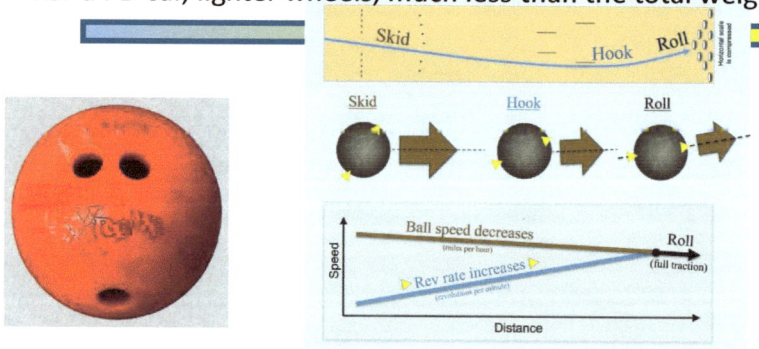

Bowling with lots of spin: The forward speed of the ball gets converted to spin as the ball stops skidding and fully rolls. As long as the ball has a hook, it is skidding and losing energy.

Bow and arrow with no spin: All the energy of motion goes into the speed of the arrow.

Each sport has a different drag. Boats have huge water drag. Bowling balls share energy with spinning.

Drag Losses in Other Sports from Air or Water Drag?

Water drag for swimmers is huge and exhausting, much worse than air drag for cars. Again, water is a 1000 times more dense than air.
Any downhill sport, such as skiing or sledding, also needs to maximize weight to air drag ratio. A heavier object, with the same frontal area, will have gravity much more than the same air drag.

"Designs should be sleek to get the least air drag."

Swimming with the drag of a life vest:
The life vest is not very sleek, although it can save your life.
Life vests are similar to the PD rectangular block, with larger air drag.

Reduce water drag by keeping your head in line with your body:
Trained swimmers can minimize drag and effort by using the right straight body.
A sleek body profile is similar to the low profile PD car.

Water mammals like dolphins live in the water and know how to just let the water flow around them:
If your life is to swim in the high drag water, you better have the streamlined right body shape.
Dolphins are beyond any PD car, where the dolphins actively use water currents to help them.

Skiing downhill using gravity:
A low air drag with a lower tuck is certain to help win races, along with a little extra weight to plow through the air.
Maybe a backpack with 20 extra pounds would help. PD cars are not allowed to pack on 20 extra ounces and dwarf any air drag.

Luge and Bobsledding downhill using gravity:
Like skiing, a lower profile is certain to help win races, along with a large body weight, if body weight is not regulated. Heavy people will plow through the air easier and have a natural advantage. Their wider body size doesn't matter for bobsledding, when athletes can tuck inside the sled. Sleds and people should be as heavy as possible, to dwarf air drag.
PD rules prohibit this extra weight advantage, by limiting the weight to 5 ounces.

Pinewood Derby shares a lot a design issues that are present in other sports, like water or air drag.

5.1 Trick 1: Drive Straight Down Monorail: Align wheels and avoid monorail drag

"I'm feeling a bit woozy ...please straighten me out."

The rubbing and drag against the monorail is the major reason Pinewood Derby cars slow down. The car that does not go straight will keep rubbing its inside edge of wheel against the monorail, just like pushing a brake pad. For any hope of competing against other cars, the car needs to go straight.

Without turning and rubbing, the car can preserve the gravitational energy from rolling down the ramp. Monorail drag can slow a car down by over 10 car distance at the PD races.

Mis-aligned car rubbing against monorail, bad

Go straight: Align wheels

This monorail drag, if your car does not go straight, is the largest source of drag to slow your car down.

It is best to completely remove monorail drag by making the car go straight.

Straight travel ...Very, very, most important ...Separate the Jedi knights from the Padawans (for Star Wars fans).

http://www.killmydaynow.com/2010/10/very-strange-car-crash-5-pics.html/
https://www.cnn.com/videos/us/2018/05/11/truck-hits-light-poles-on-highway-wisconsin-orig.cnn

Fun real world comparison

Crash: Monorail drag is not something you normally experience in a car. It is more like an **accident** you experience in a car.

Car crash on divider

Does anyone think of driving along the highway in a people car and skidding along the cement divide or barrier? That misfortune is tough driving and slow going ...you bet.

Car crash on divider: Touch monorail and get lots of drag to slow your car down, from friction.

Trick 1 🚗 Drive Straight and stop rubbing the monorail.

Monorail drag is mostly unheard of in regular cars, but boy is it important for the Pinewood Derby races.

Monorail Drag is Rough Sliding Like Brakes, Not Rolling

Monorail Drag: Most of the friction and drag that slows down the car is due to the plastic wheel pushing against the wood or metal monorail, if the car doesn't roll straight and keeps rubbing.

Monorail drag is just like applying a car brake. In a car, brake pads push against the disks inside the wheel. In a PD car, the wheel is pushing itself against the side of the monorail.

"I want to go straight, and not feel the burn of my wheels against the monorail!"

Rollerblades also using rubbing to slow the rider down.

Soap box cars push some stick or rubber against the ground, or against the wheel like it is a disk brake.

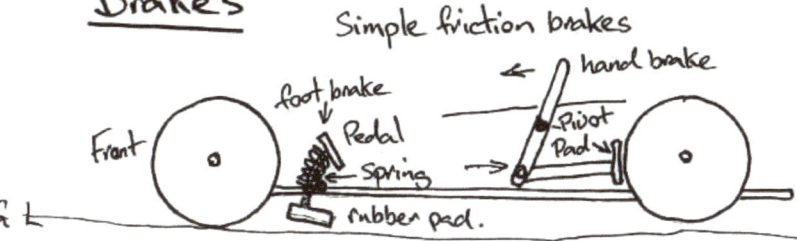

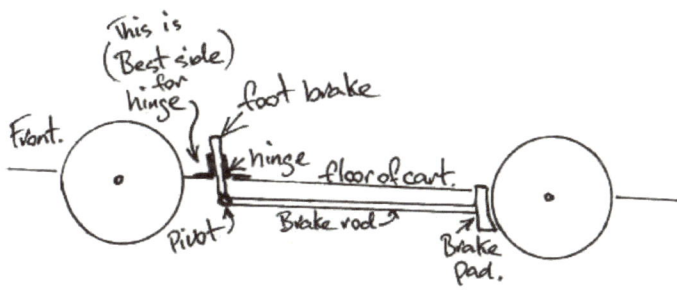

Car disk brakes.
Again, all the energy of forward motion goes into the pad and rotor as heat.

Rubber stopper on rollerblades creates friction with the ground and stops the rolling, when the skater deliberately drags the rubber.

Soap box cars use friction to slow down, either by a rubber pad against the ground or a rubber pad against the wheel. The rubber pad can be controlled by a foot pedal or a hand lever.

Bicycle disk brakes.
Disk brakes on bicycles are better than rim brakes because water and dirt does not get there.

Rubbing against a wheel is used in car brakes, by rubbing against the disks. The disks heat up and air flow needs to cool them down.

Sideways Turn Needs a Sideways Force

For Pinewood Derby, the monorail pushes back against the car to keep it going straight. This force is the pushback from the monorail and causes friction and drag, which dominates PD race times.

Anytime something turns, there must be a force pushing it to the side. The examples below, including the Pinewood Derby car, all need a force pulling them around a circle, pulling them sideways, if they are turning. The turning force is sideways to the velocity direction.

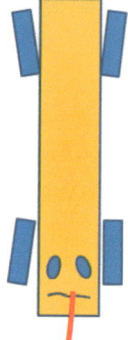

"I push to the side, and the mono-rail pushes back the same, equal and opposite."

Stopping Pinewood Derby car from turning

Force to turn equals pushback from monorail, causing friction, if the monorail is rough or the plastic wheel is rough.

Monorail Drag

When the Pinewood Derby car turns, it is pushing itself sideways.
The monorail needs to push back with the same force, to keep the car going straight.
- Drag force is the sideways force times the coefficient of friction.
- Wood is rougher than metal, and has more friction and drag force.

Turning around a corner in a car

Those tires need to push sideways on the road to turn the car.
A friction with the road is necessary to turn. On ice even turning the wheels won't turn the car.

When you are in a car going around a turn fast, can you feel the force push you sideways? Some cars even tilt around a turn, as they put more weight on the outside tires.

Ball on a string

Pulling on rope to keep ball going in circle.

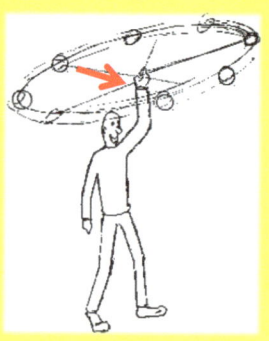

When twirling a ball overhead, the stress can get exhausting constantly pulling the ball in.
When you are pulling the ball in, that means the ball is pulling out.

Earth going around Sun

Thank you gravity, or we would be flung away from the Sun.

Gravity is pulling the Earth back in to orbit the Sun, even though the Earth wants to go straight.
Because the radius of Earth's orbit is so huge, the acceleration inwards is very small. The force is huge, but so is the mass of the Earth.

Spinning rides at the fair

Even at the fair, you exploit this sideways force.

The spinning keeps you glued to the wall of the carnival ride. You can even not touch the floor with your feet, or shift so your body is up-side-down.

You can feel a sideways force every time you turn a corner in a car. Unfortunately, for Pinewood Derby cars there is a monorail that is stopping the turn and causing drag, due to the rubbing.

Monorail Friction: Metal versus Wood Track

Metal versus Wood Track: Wood tracks are just rougher, and rubbing friction with the monorail will be larger. Our local pack track was wood, built decades before, but the regionals used a metal track.

Again, drag against the monorail is the largest effect to slow the car down. The car only has the limited energy from falling in gravity down the ramp, and once it loses that energy it is done.

Metal Track, less drag

- Less friction with metal monorail.
- Metal tracks have very little 'grab' between the wheels and the road. The mis-aligned wheels will easily slide sideways and slide against the monorail without much friction.

4 lane metal track

Metal track with a long straight-away section along the flat ground where monorail drag slows down the car.

Wood Track, more drag

- More friction with wood monorail.
- Wood tracks have much more roughness and 'grab'. The mis-aligned wheels, causing turning, will slow the car down much faster against wood than metal tracks.

4 lane wooden track, where rough wood monorail can create even more drag.

'Smooth metal tracks and monorails won't cause as much drag as rough wooden tracks, even if the car keeps turning into the monorail.'

Measuring drag on a gentle ramp:
- Place weights pulling car forward, at a crawl, no acceleration. The drag force equals the weight.
- Tilt the ramp until the car starts moving forward, slowly, without accelerating. The drag force equals the known gravity force at that tilt angle.

Lots of rubbing with mis-aligned wheels

Surface friction examples

Different materials slide over each other easier. Plastic over metal is slick with low friction, but plastic over wood just grabs.

Carpet slides at carnival

Hands-on friction race at museum

At Boston Museum of Science, build cars with different fabrics sliding over metal: cotton, nylon, and others.

If you want really high friction, use sandpaper sliding on wood

---

Cars that want to turn have more trouble with rubbing friction on a wood track. Straight-driving and turning cars are more comparable on a smooth metal track: rubbing against monorail for most any car has less friction on a metal track.

Metal versus Wood Roughness for Track

Monorail Drag: Most of the friction and drag that slows down the car is due to the plastic wheel pushing against the wood or metal monorail. This monorail drag happens when the car doesn't roll straight and keeps rubbing.

"I want to go straight, and not feel the burn of my wheels against the monorail!"

Plastic wheels against metal:
Less drag
Metal track and monorail

Plastic wheels against wood:
More drag
Wood track and monorail

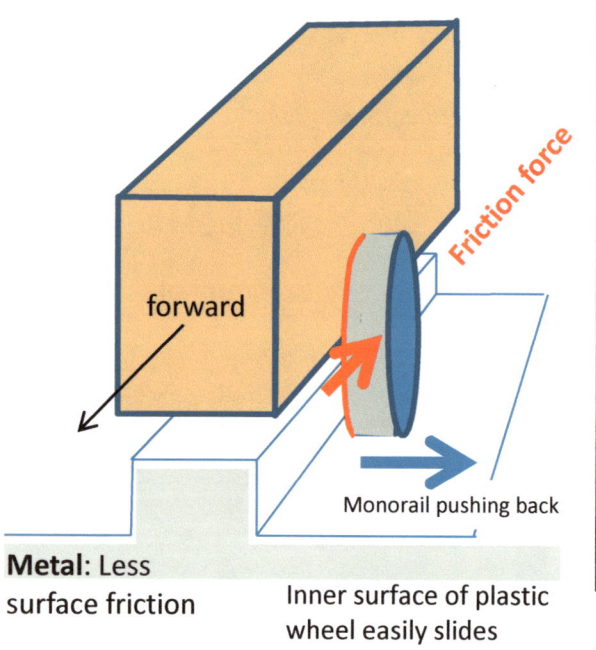

Metal: Less surface friction. Inner surface of plastic wheel easily slides sideways on smooth metal

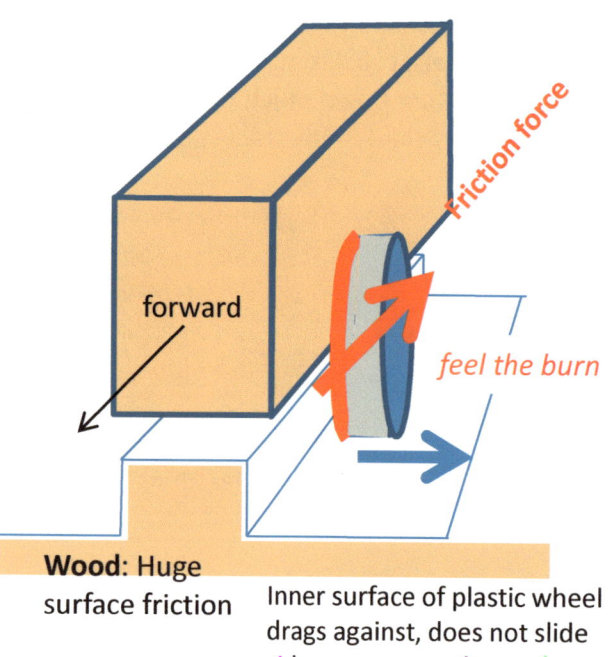

Wood: Huge surface friction. Inner surface of plastic wheel drags against, does not slide sideways on rough wood

Antique car ride with guide monorail and bumper wheels.
When at the fair ground, drive your parents nuts by banging the antique cars against the monorail, back and forth.
...except that the antique car has sideways wheels against the rail to stop the friction.

Guide monorail at fair

Wood is rougher. Wood has a lot more friction than metal, so a car that does not go straight has a larger handicap on a wood track.

Any race that is done on an aluminum track will have closer times because monorail drag is less important. Perfect wheel alignment won't matter as much, compared to a wood track. That is, race times will have larger differences on a wood track compared to a metal track.

Straight-driving cars show their speed advantage more on wood track, because they avoid rubbing against the rougher monorail.

Rolling Straight: Insert Straight Axle Nails

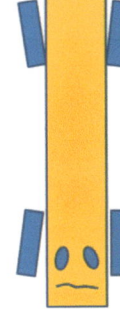

"There are many ways to straighten me out."

Here are some ways to get straight axles. In one way, you can use the provided grooves in the wood block. You can glue down the nails, or hold them in with a plastic pressure bar. Or, in another way, you can drill your own holes. To get the hole perpendicular to the block, you can use a drill press, or there are commercial drill alignment fixtures.

Alignment of the wheels is done afterwards using tapping, shims, or slightly bent nail.

Option 1: Use provided grooves and nails

Step 1: Tapping into grooves

Pre-tap the nail partly into the groove, with the nail slightly angled into the base of the groove, to keep the nail flat against the base of the groove.

Step 2 (optional): Glue nails

Glue the nails into the groove after alignment, so they are less likely to shift.

Step 2 (optional): Brace nails

Force the nails against the bottom of the groove using a commercial piece to brace the nail.

Advantage of grooves:
- simple

Disadvantage:
- nail not very secure.

Option 2: Drill your own straight holes

Step 1: Drill straight holes.

Special drill bit alignment tool.

Or use drill press with alignment block surface.

Advantage of drilled holes:
- nails straight and secure

Disadvantage:
- need to buy alignment tool, or have a drill press.

Option 3: Get a kit with an axle, instead of nails.

Step 1 with axle: Press in axle

Advantage of kit:
- simple

Disadvantage:
- none.

Drill straight carefully with tools, hammer carefully, and then glue.

Trick 1 — Drive Straight

Why Do Mis-aligned Wheels Roll to Side?

Front and Rear wheels cause different turn directions, for the same angle of the axle.
The same force would be on opposite sides of the center of mass, which causes opposite turns.

"Think of your PD car as an odd car where all the wheels can turn.
Then pick one of the hodgepodge (but logical) techniques ...such as tapping nail, rotating bent nail, shim... to straighten out the car."

Turning with dominant front wheel: If the PD car is turning toward dominant front wheel side, then need to bend this axle forward, or rotate a slightly bent nail so it is bent forward.

Turning away from dominant front wheel: If the PD car is turning away from dominant front wheel side, then need to bend this axle backward, or rotate a slightly bent nail so it is bent backward.

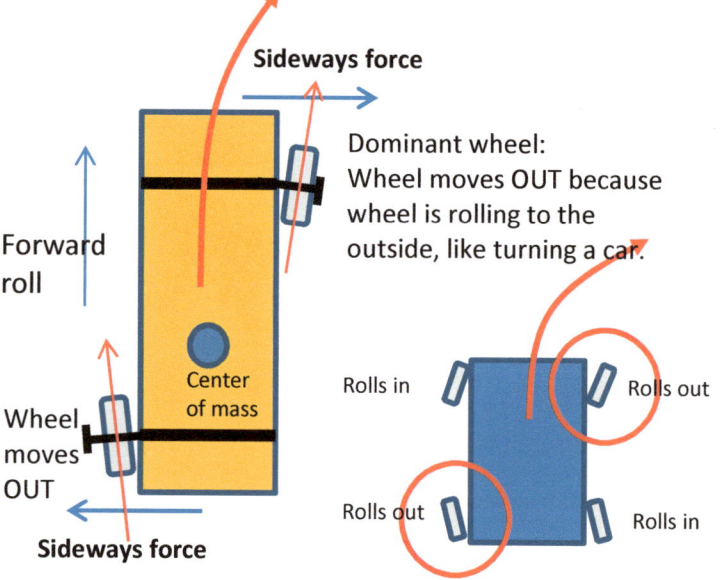

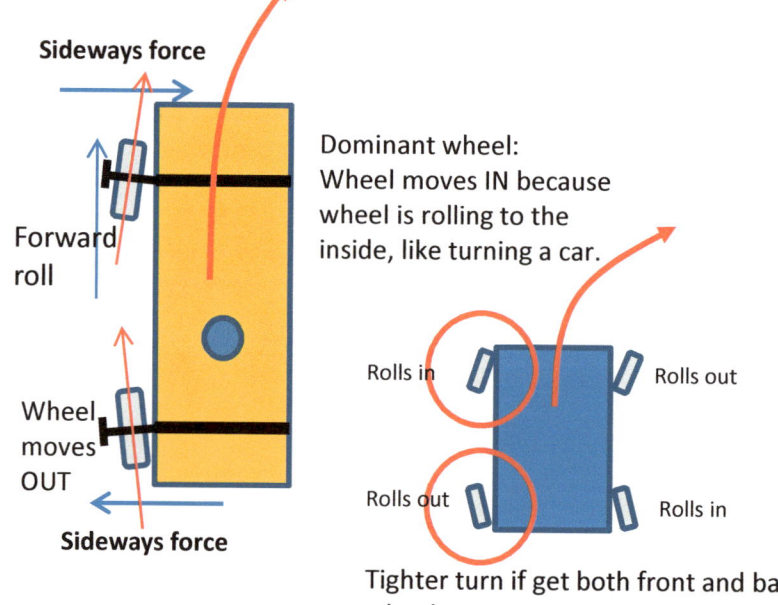

Tighter turn if get both front and back wheels to turn.

Rear wheels: If car is turning away from dominant rear wheel side, then also need to bend the rear axle forward, or rotate a bent nail so it is bent forward.

Rear wheels turn car in other direction, like a strange car with the all wheel independent steering.

The rear wheels are behind the center of mass, and the sideways force is operating behind this center of mass. For example, for the people on a see-saw, one person makes the see-saw rotate in one direction, and the other person makes the see-saw rotate in the other direction, when they are pushing off the ground to go up.

Electric Vehicle with a Four In-wheel-motor Drive and All-wheel Independent Steering:
- project completed in Dunedin, New Zealand, 2013

All wheels turn in this experimental car, for extremely sharp turning.

Trick 1 — Drive Straight

If the wood guide holes are not exactly 90 degrees, then may need to do more exact fix with shims or a bent nail...

42

Wheel Alignment when Nail is Deliberately Bent in Middle: Rotate Nail

Strangely, a very slightly bent nail can help align the wheels, by correcting for other mis-alignments. A slightly bent nail can correct for a groove that is not exactly 90 degrees. That said, there is less slipping if all the wheels are aligned and going straight.

"You can rotate the nail, if the nail is bent, until the car goes straighter and wheels don't roll in or out."

Option 1: Quick alignment: Bend nail slightly with hammer and vise and then rotate slightly bent nail

Turn bent nail until go straight

1 tap... good start

3 taps... too much

Bend nail with hammer and vise. Very touchy... 1 tap should do it with about 1 degree bend.

Let's assume the nail is slightly bent, then this trick of rotating the nail by quarter turns should work to align the wheels.
Put a mark on the nail head. Rotate the nail a quarter turn until the car goes straight, or until the wheel does not go in or out as the car is rolling, and the car goes straight.

Option 2: Use deliberately slightly-bent commercial nail on main front wheel, to help wheel alignment.

Here is an example 1.5 degree pre bent nail. (A 2.5 degree bend seems too much)

_{'Bent Pinewood Derby Axle - BSA 1.5 Degree bend with Easy Turn Screw Driver Slot - Polished Pre Bent for a Steering Axle (1 axle)'}

Option 3: Bend your own nail using custom tool

Buy an axle bender to make alignment easier, by just rotating the nail.
'Pro Axle Bender'

Real Car wheel alignment and balancing:

Wheel alignment, to stop wear on rubber

Alignment Toe-In (Caster): Place long poles along the tires, and see if parallel.

Laser alignment of wheels.
Wheels should be aligned to much better than 1 degree, to go straight and avoid wear.

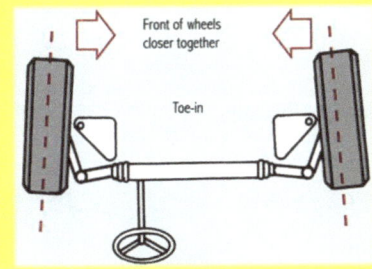

Caster alignment

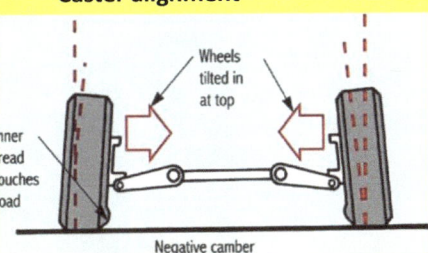

Camber alignment

Balanced regular tire, to stop vibrations

Spin the tire, and see where weights can be added to stop the tire from vibrating. This balancing machine reports where to snap on the lead weights.

Automated wheel balancing

Real Bicycle Wheel balancing:

Tighten tension in spokes to straighten wheel and stop wobble

Spoke tension to balance bicycle tires

Trick 1 — Drive Straight

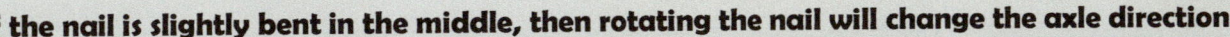

If the nail is slightly bent in the middle, then rotating the nail will change the axle direction.

Alignment: Shims, Compressing the Wood, or Nail Rotation of Bent Nail

"To get nails straight, you can use shims, tap, or rotate a slightly bent nail. The goal is for wheels to not move in or out as rolling forward, and for the car to roll straight."

Here's how to align your wheels, to change the angle of the nail axle.

Option 1: Tap nails to counter-act bend and to compress wood in your favor

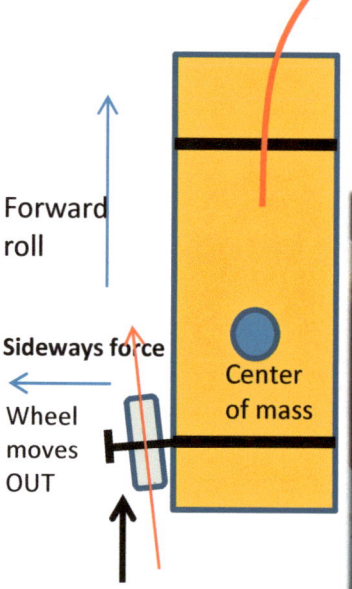

Forward roll

Sideways force

Wheel moves OUT

Center of mass

Hammer tap to straighten the axle

Tilt nail forward with a slight tap with hammer, compressing the wood. The wood is the first to give, not the fat metal nail.

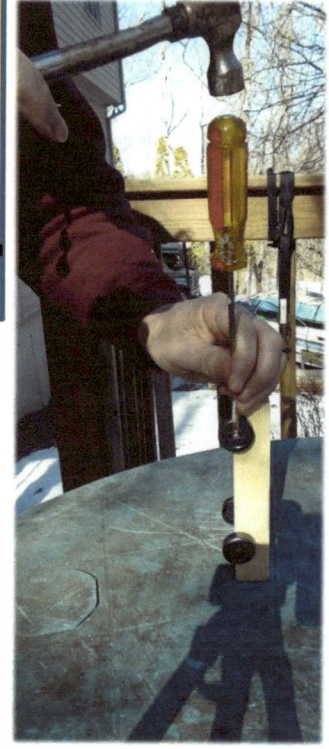

Bending nail to stop wheel moving IN or OUT.

See Appendix A for a detailed alignment process

Option 2: Rotate a slightly bent nail and see dramatic changes.

This nail orientation below rolls the straightest.

2.5 degree nail

| 0 turn of nail | 1/4 turn | 1/2 turn | 3/4 turn |

Starting roll

Coasting test at different nail rotations, to see which nail setting goes straight.

- Very dramatic changes with nail orientation: Rolling car can get large turning even over rolling 1 foot with wrong orientation of bent nail.
- A slightly less bent nail, say 1.5 degrees, is probably better than this 2.5 degree bent nail, as seen on right front wheel above, too extreme.

Trick 1 — Drive Straight

A 2.5 degree bend causes a lot of turning if not rotated correctly. A 1.5 degree bend is more reasonable. These small angles in PD cars just show that people cars need extremely good tire alignment, of 1000ths of a degree.

Car Turning and Long Wheel Base

Distance between wheel axles does matter and impacts turning. If you want quick turns and less stability, then use a short wheel base. If you want slow wide turns and lots of stability, then go long. Recall that Pinewood Derby rules allow a maximum of 4.5 inches between axles, no more.

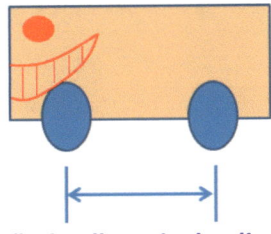

"Wheelbase is the distance between the front and rear axles, or nails."

Short cars:
Spinning in circles, with a short turn radius!

Go Karts with short shells base

Short Unicycle: The ultimate short wheel base, where you can turn on a dime.

A unicycle would be terrible for PD because a unicycle would not go straight.

Long cars:
Just go straight, with a long turn radius!

Drag race, going straight with long car

Combustion engine has huge exhaust gases
- 330 mph
- Driver in fire suit and helmet

Long Funny car: Hey, funny cars don't turn when drag racing. They accelerate straight and go ¼ mile and then pop (deploy) a parachute.

These long cars would be great for PD, but then different length cars would have different center of mass and longer cars would be most likely to win.

Cars are designed for their own purpose: Short cars are easy to turn, but you probably don't want to go fast in them. Long cars are hard to turn, and they are more stable going fast.

Turning: Compare People Car and Pinewood Derby Car Lengths

Long distances between axles will reduce the turning when one nail axle is not aligned, but the PD rules limit the maximum distance to 4.5 inches. Otherwise people would stick the axles right at the leading front edge and the very back of the car, to reduce turning.

SHORT car ———————————————→ LONG car

Shorter turn radius:
easy to turn, more monorail drag

"Help, I'm weaving like a drunk chicken!"

Pinewood Derby cars are in-between

Larger turn radius:
harder to turn, less monorail drag

"Straight as an arrow"

Real world comparisons of People cars:
Short cars turn easily, so cars are typically short.

Sedan: ~1.75:1 wheelbase ratio
- Good turning radius to maneuver in cities, and park easier.

A Standard motorcycle: This shorter wheel base is easier to turn. Again, the smallest wheel base is a unicycle.

Ratio of wheel base – that is, separation between axles – to width of axle:

Pinewood: ~2:1 wheelbase ratio
- Harder to turn than standard car (if had steering)

Pinewood Derby cars have a 2:1 ratio. Most sedans are less at 1.75:1, and SUVs are more at 3:1. So Pinewood Derby cars are designed for just going straight.

Experiment: Make a longer stretched Pinewood Derby car and see if the longer car goes faster down a monorail track, using the same weight limit of 5 ounces. The longer car should go straighter and have less drag against the monorail. Less drag and going faster means slowing down less.

Real world comparisons of People cars:
Use long cars if don't need to turn

Funny Car Dragster: ~3:1 wheelbase ratio
- Nobody turns these bad boys.

A Chopper motorcycle with forks: This longer wheel base is harder to turn than a standard motorcycle.

The Pinewood Derby rules give the maximum and minimum distance between the axles, but if you want shorter distance you would get a less stable design. Longer axles separation and the car goes straighter, with less monorail friction.

5.2 Trick 2: Axle Friction: Graphite and Smooth Nail

Those axle nails need to be smooth and friction free. How? Load the nail axle up with light lubricant before the race, and get rid of any burrs on the inside of the nail head. Then the car can preserve the gravitational energy from rolling down the ramp. Low friction axles with lubrication can improve race times by over 3 car distances.

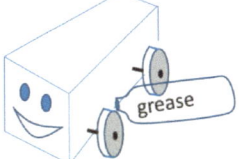
"Ooh, so cool and smooth. Thank you grease."

Make Pinewood Derby wheels spin without friction

Squeeze graphite on and spin wheel

Graphite lubrication: less nail friction

This nail friction can be a big deal if no light lubrication is used. During the race, cars visibly slow down on the straight-away, and cars get slower as the lubricant wears away from race to race.

- **Graphite powder** is used as a grease or a light lubricant. Graphite has little flakes of thin carbon film that slide over each other.
- **Graphite** is also the 'lead' in pencils, because the graphite flakes can easily slide off each other and stay on the paper.

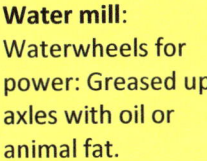

wheel — velocity — Force drag

Lubricate the smooth nail with light graphite powder
- Flaky powder works
- Light oil works
- Heavier oils will not help, because the car is too light to push through the stickiness of the oil.

Again, the best ways to go fast are to get a straight coasting car and to use light lubricant on the axles.

Real world primitive axle: Here are other ways to make wheels spin without friction in history

Greased axles on horse-drawn cart

Carts in middle ages used greased up axles, using animal fat

Wood shaft and grease

Water mill: Waterwheels for power: Greased up axles with oil or animal fat.

Wood shaft and grease

Real world primitive fire starter:
Heat from spinning stick
Same friction without grease is used to start a fire, spinning a stick... something does not add up, but this is wood on wood! Fortunately, Pinewood Derby has metal on plastic for nails and wheels, so less nail friction.

Wood on wood with high friction, no grease, to start fire.

Trick 2 — Axle Drag

Axle Friction: Remove Nail Burr on Wheel Axle

Some nails come with 'burrs' under the nail head. This is from stamping the metal down to make the head.
You need to file this burr off so it doesn't rub against the inside of the plastic wheel.

"Smooth nails and graphite work great, maybe better than sticky greased bearings for this small weight and short race."

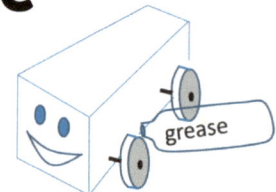

De-burr and polish the nail

Get the nail shiny smooth:
- Use a drill and sandpaper to deburr and polish the axle of the nail.
- Use huge amounts of lubricant to reduce axle drag even after the nail is polished.

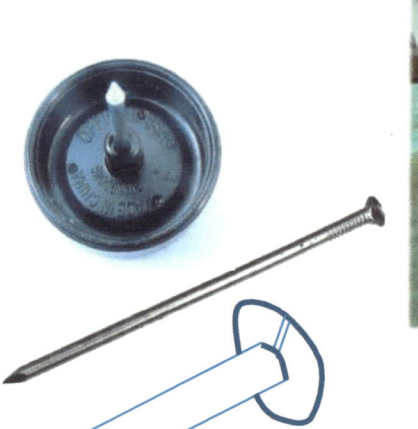

Burr (from stamping the nail head)

Spin a drill to sand the nail quicker

Remove nail burr with sandpaper or a file

Burr:
If there is a big burr on the underside of the head of the nail, then sand the burr off. Fortunately, the nails in Pinewood Derby kits are really polished and smooth these days.

Nails needs to be filed and sanded to be good axles.

Real world comparison: use roller bearings with grease instead, never a bare axle like the old days:

- **Wheel bearing**: no slipping, all rolling
  Car cylindrical bearings are deliberately cylindrical. The cylindrical bearings are tilted and can take sideways pressure from turns as well as roll forward.

Cylindrical bearings with conical shape to handle turns

- **Bicycle ball bearings**

Re-greasing bicycle bearings.

- **Rollerblade ball bearings**.

Ball bearings on skates

http://science.howstuffworks.com/transport/engines-equipment/bearing1.htm

Trick 2 — Axle Drag

Axle Friction: Add Grooves on Wheel Axle, for Graphite Reservoir

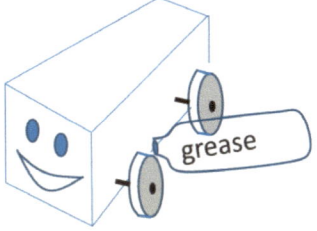

"My lubricant wears away after many races, and I need a reservoir."

File Grooves in Nail:
You can also file grooves in the nail, to store lubricant, so there is a reservoir for lubricant over many races. Without a reservoir, you can see some cars get slower and slower after repeated races, as the lubricant blows away.

Use spinning nail and file to create groves.

Grooves:
Sand in grooves around the nail, so the lubricant can have a storage home, and not be pushed away after the first race.

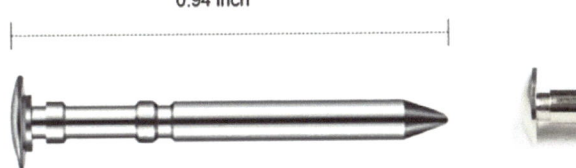 

0.94 inch

Red Dirt Derby Machined, Polished, Grooved and 2.5 Degree Bent Axles - (2 Axles) for use in Pinewood Car Racing

Examples of grooves to store the lubricant.
The graphite gets stored in grooves, for the next race.

- The Pinewood Derby can use grooves in the nails to temporarily store the lubricant, so the car has grease over many races.
- Real car bearings keep the grease enclosed using caps over the bearings.

Nail grooves can be a reservoir of extra lubricant, to last many races without a refill.

Any advantage using a ball bearing for Pinewood Derby?
—Maybe bearings have no advantage for a light weight PD car and sticky thick ball bearing grease.

Ideal bearings have no friction because everything is rolling and not sliding. However, real world bearings have heavy grease (for people cars, for roller skates) and are not ideal for the light Pinewood Derby car:

- First, the compression of the bearing grease can cause loss. Typical bearings have moving thick grease which is sticky and may give more drag than a nail with light powdered lubricant, such as graphite. Ball bearings are really meant for heavier weight where bearing grease stickiness doesn't matter.
- Second, even if there is less friction using bearings, the bearings have weight and more energy goes into angular momentum of wheel, and not into the speed of the car. To experiment, we should build a car with bearings and compare the speeds over the short track.
- Try Teflon sleeve instead of bearing?

Bearings are meant for a heavier weight:
- too big,
- too expensive,
- not effective for light weight (sticky with heavy grease),
- not allowed

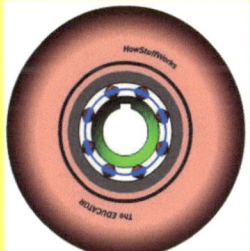

Rubber roller skate wheel with bearings.

Ball bearings are meant for heavy loads, not light loads. For light loads, the heavy grease resistance or stickiness makes bearings less useful, and the mass of the bearing just wastes energy in rotation.

Trick 2 — Axle Drag

History of Round Finish Nails

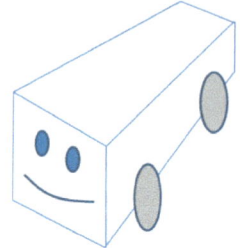

Nice round finish nails are a good axle.

Nail manufacturing has gone through a lot of improvements, in terms of cost and availability. Before 200 years ago, nails were not round. Nails were custom made and were quite expensive. An ironsmith had lots of work making nails and horse shoes.

Over 200 years ago, nails had a square cross section. The iron nail were made by hand by hammering on iron to get the carbon out. There were no machines for rolling iron into smooth circular rods.

About 200 year ago, nails could be stamped out of thick sheet metal. Stamped tapered nails actually have an advantage for building houses. These type of nails hold better and they don't split the wood as easily.

"Nails are now just cut smooth wires."

Hand made nails, in Middle Ages

Hand forged nails have been found from Roman times 2000 year ago.
Each was custom made, and the square cross section means they are not an axle for a wheel.
These nails have a long taper and are easy to drive into wood.

Old cut nail, 4 sided taper

Stamped cut nails, in Colonial Era

A tapered nail pushes the wood grains downward, which resists the nail pulling back out.
Also, there is only a taper in one of the dimensions. So the nail can be aligned with the grains to avoid splitting.

Old stamped nail, 2 sided taper

Finish nails, or wire nails, with modern mass production

Wire nails are formed from wire. Usually coils of wire are drawn through a series of dies to reach a specific diameter, then cut into short rods that are then formed into nails.

Modern wire nail, no taper but inexpensive

Wire nail making machine: A fat round wire enters the machine, and the head gets hammered and the pointed end gets cut.

*Stumpy Numbs Woodworking Journal

Smooth round finish nails are quite convenient as an axle for the Pinewood Derby car. Finish nails started out as long wires that were cut, so of course finish nails are round.

Skinny Axles Have Less Drag

"Bearings should be small, but still strong enough to hold the weight and handle bumps."

A skinny wheel hole for the nail axle allows less drag.

There is always some friction at the axle. The inner plastic of the wheel is actually sliding over the metal nail. Fortunately, that sliding is very slow because the diameter of the wheel hole is very small. So there is not much drag or torque from this drag.

Some friction is also true with car bearings, even with rolling balls inside the bearings and not sliding surfaces. It is better to have smaller diameter bearings, because the bearings are not actually rolling or sliding very far as the outer rim of the wheel rolls a great distance.

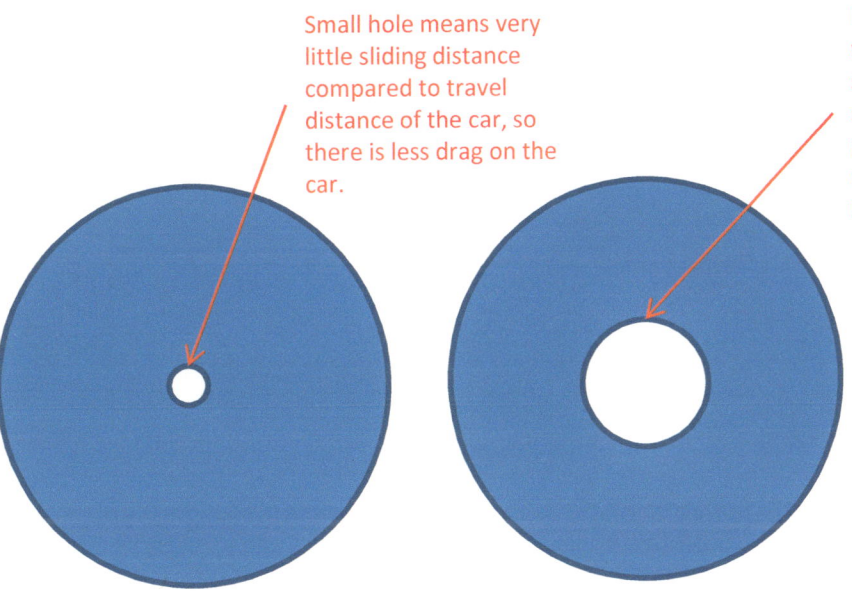

Small hole means very little sliding distance compared to travel distance of the car, so there is less drag on the car.

Inner plastic of large hole would travel almost the same distance as the car, so drag between the plastic and the large diameter nail would be impactful.

Tiny hole for wheel   **Big hole for wheel**

The tiny hole does not actually roll very far as the car goes along. The distance travelled by the inner whole is just the ratio of the diameters, from the inner hole to the outer wheel diameter.

Bearing diameter does matter. Smaller bearing move less and will enable less car drag, if the bearings can handle the load and bumps.

Bearing diameter in cars and bicycles?
Bearings are all around us, and bearings try to be small.

Car wheel bearing need to be large enough to be strong, but not so large that any bearing friction has a large impact on the car drag.

Bicycle wheels have a very small bearing compared to the wheel diameter. That means the bicycle should coast very well, where any bearing friction does not impact the overall bike friction much. A small bearing is possible because the strength of the bearing does not need to be huge.

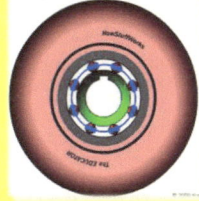

Roller skates actually have a relatively large bearing diameter compared to the wheel diameter. That large diameter is less expensive and stronger, but causes any bearing drag to have more impact.

5.3 Trick 3: Avoid Air Drag!

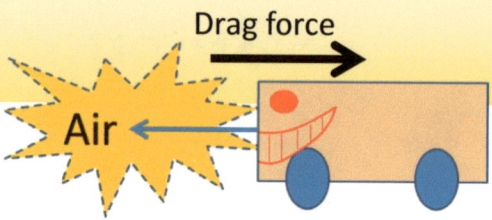
Drag force

Make the car slim, sleek, and low profile, to reduce air drag. Then your car is able to go as fast as it can, and preserve the gravitational energy from rolling down the ramp. Air drag from a tall car can cost an extra 1 car length at the races.

"Want air drag? Go tall and wide, hitting air like a belly flop on water."

Air drag comparisons on other moving transportation:

Air drag is real, people!
- A car, on the highway, pushes against the air drag force, the same weight as a person standing on top of you, wasting fuel.
- A truck, on the highway, pushes against the air drag force, the same weight as a few refrigerators laying on top of you, wasting fuel.

Feel the air drag out the car window.

First, make a slim car to have small frontal area and less air drag.

Second, air drag is not all about being slim. The right shape, even after the car is slim, will help the air flow around the car without creating turbulence. Turbulence creates a little lower air pressure in the back, which pulls the car back and causes a drag force.

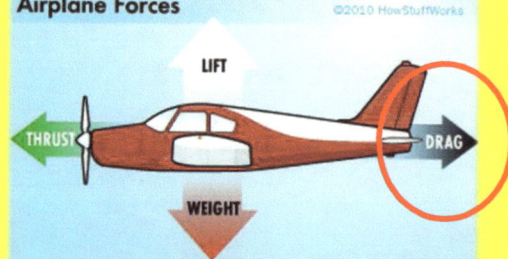
Airplane engines fight air drag when flying straight and level.

Where does most of the fuel of flying go? Air drag. Partly because of air drag, jet planes go up high where the air is thinner and the air drag is less. Planes go high up also to fly above storms and avoid noise pollution on ground.
Lower area and lower drag when in crouch.

Slim car, though not so sleek

Experiment: Make heavier Pinewood Derby car, greater than weight limit of 5 ounces, with same shape and same frontal area. Air drag force is now a smaller fraction of the gravitation force, so the car should not be slowed down by air drag as much.

Slimmer cars will lose less of the power from air drag. PD cars are stuck with limited power from gravity, so don't waste that energy.

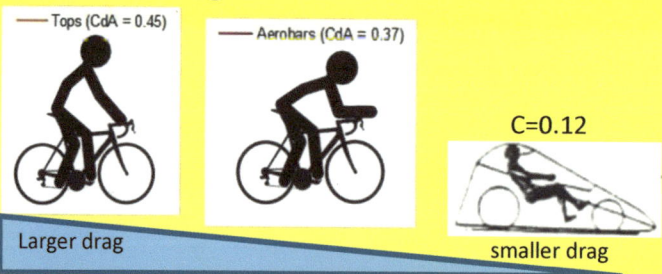
Bicycling straight and level is fighting air drag.

What stops you going more than 30 mph on your bicycle coasting down a hill? → Air drag.
(and maybe a wobbly wheel and a healthy fear of crashing)

Trick 3 🚗 Air Drag slim car

Air Drag Takes Away Speed

Air drag is something that always needs to be considered for any moving design. It can be good, it can be bad. For Pinewood Derby, air drag is bad. For parachutes, air drag is good.
Let's look at a major benefit of air. For breathing, air in general is good.

"Air drag is good for braking and air is good for flying.
Air drag is not so good for traveling along the highway. On a straight-away, air drag is the main reason to use an engine to keep the speed."

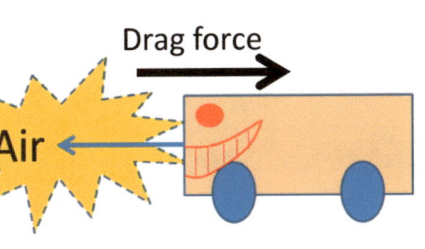

Slim design, like streamlined funny cars before they deploy a parachute: Less air drag and less area

Funny car drag race

Wide design, like funny cars with deployed parachutes: huge air drag and huge area

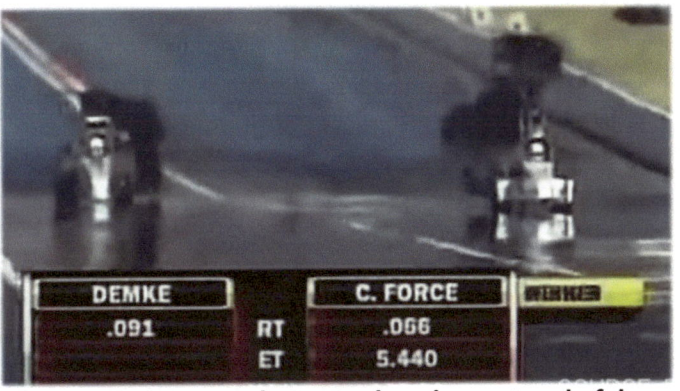

Funny cars use parachutes to slow down at end of drag race

Bicycle wrapped for low air drag

Slim car with no drag from mirrors and rear wheels.

With an aerodynamic shape, there is less pushing the air to the side and less turbulence due to any broken air stream behind the car.
The wrapped bicycle above has smooth air flow with no broken air stream.
https://en.wikipedia.org/wiki/Automobile_drag_coefficient

Bicycle with no air drag reduction, but fun to feel wind on your face
To reduce air drag, people get covered spokes and teardrop posts. After all that, don't sit bolt upright with huge body area facing the wind.

Car styles worry about air drag all the time. Unfortunately, bicycles or motorcycles typically don't.

Push Air Aside and Lose Energy

Air has mass, and needs to be pushed out of the way to allow the car to barrel though. If you want to push away less air, then make your car slimmer and lower profile. A lower air drag car can buy you 1 car length at the finish line.

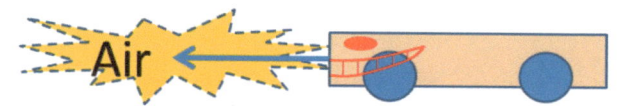

"Slim is faster. Just keep the same 5 ounces using extra weights, and you'll avoid large air drag and go faster."

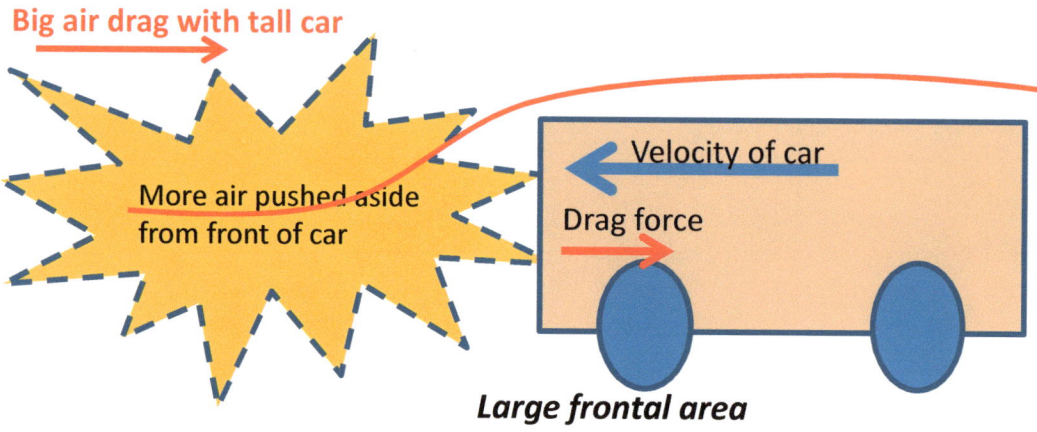

There is low pressure in back pulling car backward, like a little vacuum. Turbulence creates the low pressure.

Simple block, ready to buy: High air drag

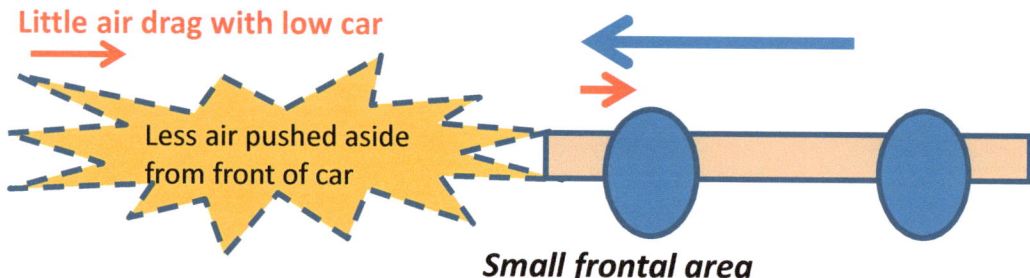

There is less air being pushed around using a lower profile car.

Low profile car at max weight, ready to buy: Low air drag

This low profile car, with most of the lighter wood cut away, uses lead weights to get back up to 5 ounces, and still keep the low air drag and small frontal area.

Air needs to get out of the way.

To lower air drag, it is first more important to reduce the frontal area of the car, and then afterwards to streamline or contour the shape to reduce drag coefficient. For most standard range of drag coefficients, it is the frontal area that matters most.

https://en.wikipedia.org/wiki/Automobile_drag_coefficient

Clearly, even from cartoon pictures, a lower profile car has less air drag.

Front Pressure and Back Vacuum

One way to look at air drag is to think you are creating a slight vacuum behind the car. This rear vacuum is lower pressure and pulls back on the car.

"The physical front area is most responsible for the drag, but you can also use a more contoured shape."

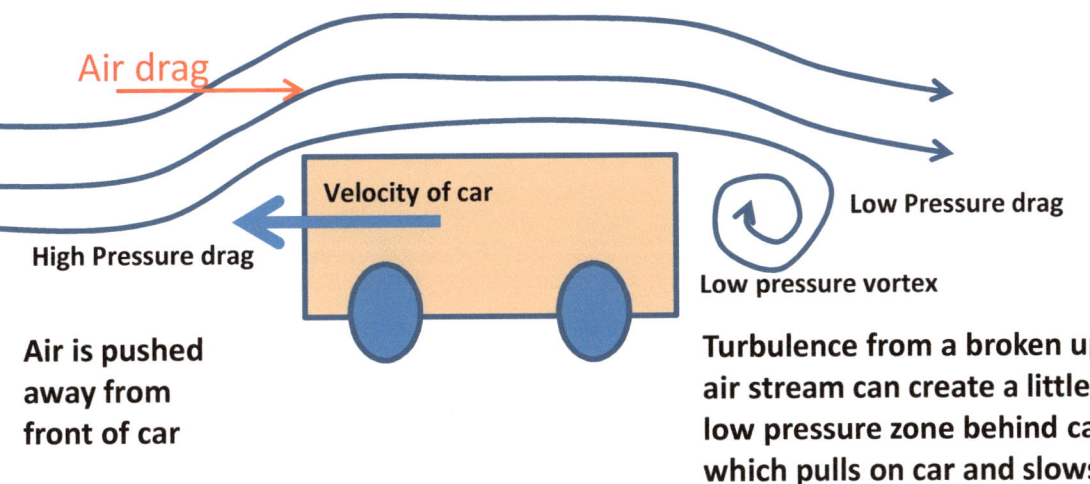

Air drag

Velocity of car

High Pressure drag

Low Pressure drag

Low pressure vortex

Air is pushed away from front of car

Turbulence from a broken up air stream can create a little low pressure zone behind car which pulls on car and slows car down.

Real world:
Air drag is a *good* thing for many designs:
Parachutes: huge air drag because of huge area.

Funny car at end of race

Shorter landing of jet airplane

Lego man still needs a larger parachute area.

Maximum front area looking down car.

More frontal area has more air drag

Real world:
Air drag is a *bad* thing for many designs:
Engine power:
- Efficient cars, especially with under-powered engines. Most of the gasoline is fighting air drag at highway speeds.
- Trucks, where drag causes poor miles per gallon and gas and dollars
- Bullet trains, which go over 150 mph
- Hyperloop concepts are even faster than bullet trains, and would have more drag without air removal. Prototype designs can suck out air from front of train using a turbofan jet, or create a partial vacuum in the tube.

Human power:
- Bicycle tires can cover up spokes, but the human rider has much more drag than the tires, so crouch down if you please.
- 'Albatross' is a human powered airplane that crossed the English Channel.

Shock waves

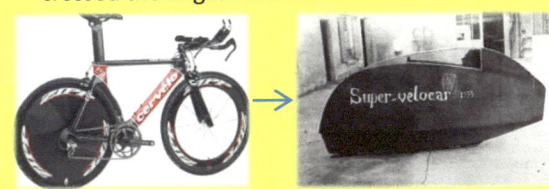
Streamlined spokes   Streamlined air flow

Human powered airplane

Faster speeds have dramatically larger air drag.

Backyard Air Drag 'Rolling' Measurements

Air Drag Test 1: Using a leaf blower, blow wind across the car from standstill, and see how far back the car rolls. If the cars are the same weight, then the distance the cars roll back should tell which car has more air drag.

"Everything moving has air drag. Airplanes use air to stay up, but also have air drag."

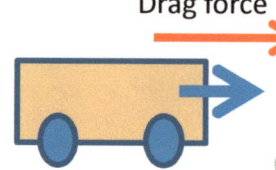

Wind speed: 5 meters per second.

Release two cars at same time, and video record them as they roll back. The car with more air drag will roll back quicker.

The taller block car moved back farther due to air drag.

Cars on flat

Leaf blower

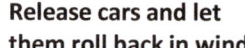

Release cars and let them roll back in wind.

Leaf blower, perched to aim wind at cars

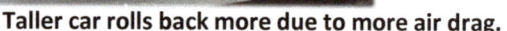

Taller car rolls back more due to more air drag.
Time to roll back is about 1 second, from (34 frame / 30 frames per sec) * (1sec)

Use a wind tunnel to measure air drag, or ...

Car designers can measure the air drag in a wind tunnel.

Wind tunnels measure drag and turbulence.

You can test the drag on your people car, using a video camera and the car coasting in neutral.

Just find an empty straight-away on a highway, with less than 1 degree gradient, and coast fast.

Car coasting in neutral on straight-away.

Record distance and speed as coast and slowing down.

To measure air drag, put your video camera on the speedometer and record how quickly the car slows down. Knowing the weight of the car, and the frontal area of the car, you can determine the drag coefficient, or an upper limit. Rolling resistance (tire compression) also contributes. People have tested drag on police lights this way. Screw on different police lights, and compare how fast the police car slows down while coasting in neutral on the same track.

Not the most reliable of measurements, but all the shorter cars seemed to get pushed back at same rate.

Backyard Air Drag 'Rolling' Measurements

Air Drag Test 1: Using a leaf blower, blow wind across the car from standstill, and see how far back the car rolls. If the cars are the same weight, then the distance the cars roll back should tell which car has more air drag.

"Everything moving has air drag. Airplanes use air to stay up, but also have air drag."

wind → Drag force → Pushback Velocity of car

Wind speed: 5 meters per second.

Cars on flat. Leaf blower, perched to aim wind at cars.

Release two cars at same time, and video record them as they roll back. The car with more drag will roll back quicker.

The taller block car moved back farther due to air drag.

Release cars and let them roll back in wind.

Taller car rolls back more due to more air drag.
Time to roll back is about 1 second, from (34 frame / 30 frames per sec) * (1sec)

You can test the drag on your people car, using a video camera and the car coasting in neutral.

Just find an empty straight-away, with less than 1 degree gradient, and coast fast where air drag dominates.
If you coast slow then tire resistance dominates.

Car coasting in neutral on straight-away.

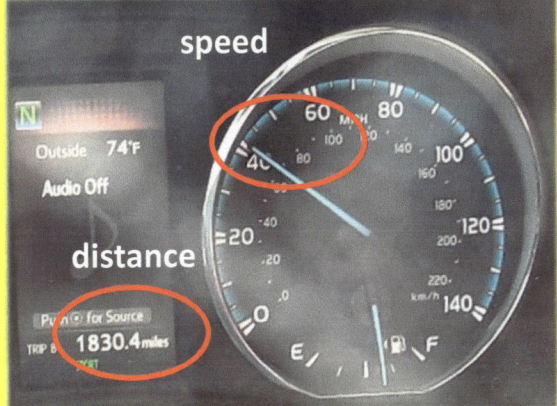

Record distance and speed as coast and slowing down.

To measure air drag, put your video camera on the speedometer and record how quickly the car slows down. Record the speed at each 0.1 miles for a coasting car, to get the coefficient of air drag.
Knowing the weight of the car, and the frontal area of the car, you can determine the drag coefficient, or an upper limit. Both air drag and rolling resistance (tire compression) contribute.
People have tested drag on police lights this way. Screw on different police lights, and compare how fast the police car slows down while coasting in neutral on the same track.
Force = mass*acceleration
= mass*change in velocity over change in time.

The roll back is not the most reliable of measurements, but all the shorter cars seemed to get pushed back at same rate.

5.4 Trick 4: Weight in Back Gives More Speed

There is more time to pick up speed when anything starts higher up. Look at ski jumps. Look at unfortunate people who fall from great heights.
Yes, all the energy in Pinewood Derby comes from gravitational energy, which is just the starting height. More starting height means faster speed at the bottom.

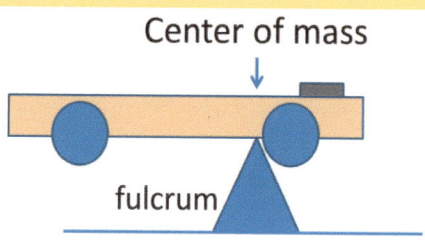

"Make me heavy in the rear end, and I'll go faster."

How does this more height relate a Pinewood Derby car, where all the cars start at the same height? Get the most mass toward the back, and the center of mass of the PD car is higher at the starting line on a sloped track.

Admittedly, getting an extra 1 inch of starting height for the center of mass is only a slight advantage (beyond the starting 3 or 4 feet starting height), but it is something. The car can be 1 car length ahead just due to the extra starting height.

When the ski jumper starts on a higher ramp, then they have more speed at the bottom and jump farther.

Two cars perched to get released.
When the most mass of the PD car is in the back, then the car effectively is at a higher starting height, even though the front end is at the same location.

Get an extra inch of height of center of mass by starting on a ramp! This height and extra energy is just one of the many little tweaks that are needed to be the top dog contender.

Trick 4: Weight in Back Gives More Speed

The race cars start at a steep angle on the starting ramp, for Pinewood Derby without an engine. That means that a car with a center of mass that is toward the rear of the car is effectively higher up the ramp.
Yes, the starting height is not just where the wheels are. The starting height from the gravitational energy point of view is where the center of mass is, higher up the ramp.

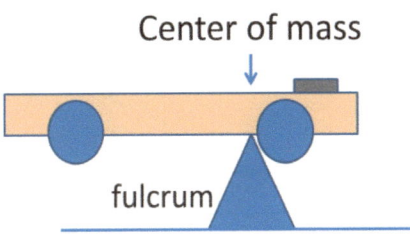

"Make me heavy in the rear end, and I'll go faster."

This car has its mass distributed along the car, so the center of mass is more forward and has less starting height.
Both cars are at the upper limit of weight at 5 ounces.

This car has some lead weight screwed in the back, and there is more wood in the back.

Extra 1 inch height of center of mass using weight in back.

Two cars perched to get released.

Front and rear axles: Some cars have an axle closer to back end, so can get center of mass closer to back, and have a sharper nose.

All else being equal, the car with weight in the back should win by 1 car length.

Weight in back to reach higher speed:
Yes, there is a reason that Pinewood Derby kits put a large weight in the back. There's more gravitational energy when starting on a slope.

Fast car premade to order, with all the right fast tweaks.

Other examples of the PD cars on the starting ramp:
All the fronts are lined up, but the actual weight location or center of mass location of the cars are different.

Steep 6-lane Launch

Steep Launch:
Really steep, so weight in back helps even more.

Get an extra inch of height of center of mass by starting on a ramp! This height and extra energy is just one of the many little tweaks that are needed to be the top dog contender.

Weight in Back Gives More Height and Bounce

Lead weights are your friend. The more you place the weights near the back on a sloped ramp, the higher the car is in terms of center of mass and the more energy it has from gravity.
Lead weights are good because the lead is soft and can be clipped to the exact 5 ounce weight.

*"I feel strangely balanced, for best speed.
Separate the Jedi Masters from the Jedi Knights, small improvements do, with even 1 car length advantage."*

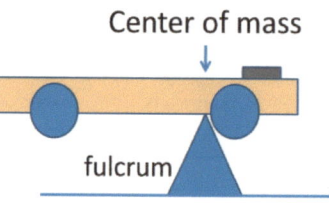

Anything starting up higher up will have more energy.

Two back weighted cars, still 5 ounces   **Back weighted and low profile car**

Bouncy ball height demonstration

Higher balls hit the ground with faster speed.

The bouncy ball bounces back up, with a higher bounce if start higher.

Weight goes in the back: Still need to keep the center of mass 1 inch in front of the back wheel, not behind the rear wheel, or the car will tilt back.

Trick 4 — Weight in Back

Weight is higher above ground when in back near the rear wheels, on the tilted starting block ramp.

Because car starts out at an angle, this trick of weight in back works. The total change in height of the center of mass at the bottom of ramp is larger when the weight is in the back. Gravitational energy is maximized.

When the front wheels touch the flat straight-away at bottom turn, the car is still falling if the weight is in the back. If the weight is in the front, the car is done falling.

As a 'thought experiment', if car started out horizontal, this trick of weight in back does not work. The change in height of the center of mass would not depend on where the center of mass lies along the car.

You can buy lead weights, in many shapes, at hobby stores. Just screw them on, but don't exceed 5 ounces.

Weight in Back of Car Gives More Height

Anything falling or rolling down a hill will be going faster the higher up on the hill it starts. If you are skiing down hill and you want to jump farther off a ramp, you start further up the hill.

Watch ski jumping in the Olympics. Alternatively, look down when you jump off a high diving board. You know you'll hit the water faster when the board is higher.

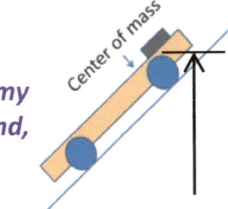

"I'm higher, my heavy rear end, that is."

Lower starting height: slower velocity at bottom

Start lower | Speed shown by distance between two frames in time | Rebound height

Higher starting height: faster velocity at bottom

Start higher | Faster Speed shown by larger distance between two frames in time | Higher rebound height

As anyone who's bounced a ball will know, the higher you drop the ball, the faster the ball will hit the ground, and the higher the ball will bounce back up (unless the ball goes smoosh).

The Pinewood Derby car starts on a slope, so if we can move the center of mass of the car toward the rear wheels we effectively raised the car about an inch extra. This gives more speed at the bottom, just like a ski jumper.

Trick 4 — Weight in Back

Let go of ball at top.

Weight in back, with higher center of mass. Same as dropping ball from higher up.

Center of mass over rear wheel.

Speed at bottom: Gravitation energy went to Kinetic energy:

Gravitational Energy = mass*gravity*height

Speeding up from gravity

Fastest speed at bottom.

Velocity at bottom is faster when start the car's center of mass at more height.

Coasting, and slowing down from drag.

A basketball, or superball, or beach ball shows that anything dropped from a large height has more energy at the bottom.

5.5 Trick 5: Lighter Wheels and Minor Stolen Energy

Spinning wheels have energy, just from the spin.
Look at cars. It actually takes more power to accelerate forward when the wheels are larger and heavier because energy is getting put into the spin, not just into the forward motion.
Look at flywheels or gyroscopes. These spinning disks are not going anywhere, but they obviously have energy.

Newer Pinewood Derby cars already have lighter wheels, so scouts don't need to drill holes in the wheels to lighten them, but you can.

Drilling out the side walls of the plastic wheels should improve the race time by up to ½ a car length.

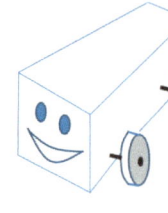

"What I really want are ice and skates, so no energy goes into spinning my wheels."

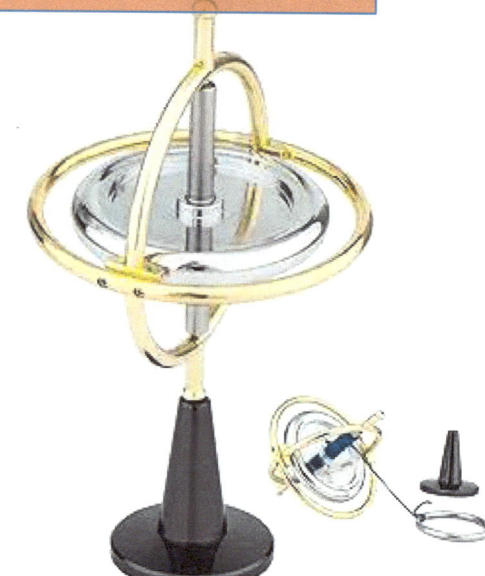

The spinning wheels take energy. For PD cars, this energy is not going into the forwards speed of the car. The PD car has a fixed gravitational energy, and we want to have the least possible energy go into the spinning wheels, so lighter wheels are a good idea.

Spinning disks have energy. Look at toy gyroscopes, that stay balanced because of the spin.

Spinning disks have energy. Look at the Segway for personal transportation, that stays balanced because of the spin.
There is a heavy, fast spinning disk inside the bottom of the Segway.

Less wheel weight reduces unwanted rotational energy. Also, raising one of the wheels, so it does not rotate, also reduces rotational energy. Of course, you still need to make sure the 3-wheel car goes straight.

Trick 5: Lighter Wheels and Minor Stolen Energy

Pinewood Derby cars want lighter wheels:
Wheels are great. They allow the car to roll forward. But wheels have spin energy as well as forward motion energy. About 4% of that precious gravitational energy is now trapped in the rotation of the wheels, instead of in the forward speed of the car.
Newer Pinewood Derby cars have these lighter wheels, so scouts don't need to drill holes in the wheels to lighten them, but you can.

"What I really want are ice and skates, so no energy goes into spinning my wheels."

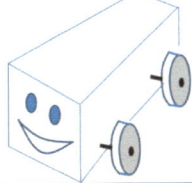

Option 1: drill holes in wheel side walls to lighten wheels

Holes drilled in the sides of the wheels will reduce their rotation energy and mass.

Newer wheels:
- 2.8 gram with weight away from rim
- 4% of energy

Modified wheels with holes

Option 2: lift one wheel off the track so it won't spin

3 wheels are actually slightly better than 4 wheels for Pinewood Derby. There is less energy lost in the spin of a 4th wheel, although going straight with 3 wheels is more of a challenge.
Of course, if the raised wheel hits the ramp every 5 feet or so and needs to be re-spun, then more energy is lost in the repetitive re-spinning.

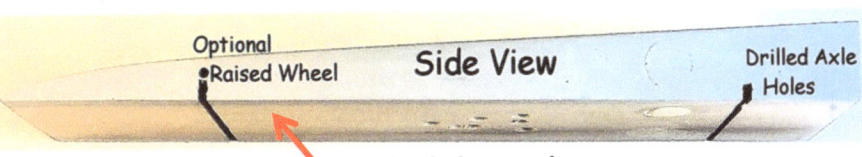

3 wheeled car wedge

Optional raised wheel, so only 3 wheels are spinning and less energy goes to rotational energy, by PineWood Pro.
http://www.winderby.com/m02_040829.html

Less wheel weight reduces unwanted rotational energy. Also, raising one of the wheels, so it does not rotate, also reduces rotational energy. Of course, you still need to make sure the 3-wheel car goes straight.

Sometimes heavy tires are great:
How much energy goes into the wheels, and not into forward motion?

Monster truck in the air with huge tires

Monster trucks have huge tires for jumping.

>400 lb tires each

Monster Trucks : 10% rotation energy
Huge tires are good for jumping like on a trampoline.

Motorcycles use rotational energy as a critical part of balance:

Typical motorcycle with tires a large part of the weight

>60 lb of rotating gear per tire

Motorcycle : 10% rotation energy
Heavy spinning tires are good for balancing or stability of the motorcycle from angular momentum when have a 2 wheeled vehicle.

Balancing without effort using spin

Segways have spinning internal flywheels and outer wheel, and acceleration is not the point.

Segway : >50% rotation energy

Balancing is the point, not cruising speed. Spinning flywheels are good for balancing upright from angular momentum.

Lighter Wheels Take Away Less Forward Energy

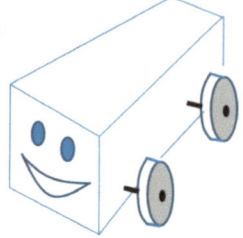

Lighter new wheels or older heavy wheels?
As said before, the wheels are there, and of course they need to be there, but the rotation energy of the wheels takes away from the forward motion. For Pinewood Derby cars, the wheel is taking precious energy away from our 4 feet height of gravitation energy, and this spinning energy is not going into forward motion of the car.

"But I wanted big fat wheels with huge springs ...with small wheels now I'll feel all those bumps."

Compare to hula-hoops. Which has more energy, a plastic hollow spinning hula-hoop or a hula-hoop filled with sand? The one with sand has more mass and more energy.

2.8 gram

Motorcycles want heavier tires for balance:
If this were a two-wheeler motorcycle pinewood derby, we wouldn't be having this conversation about reducing wheel weight to go faster. We'd want to increase wheel weight to get better balance.
Here are two-wheeler examples where the angular momentum and weight of the tire do help balance:
- Bicycles depend on spinning tires and angular momentum for balance;
- Motorcycles depend on heavy tires for balance even more because motorcycles are heavy and are faster.

Avoid the older heavy wheels:
More mass at rim, heavier, so more energy goes to rotation: about 6%.

Newer light wheels:
- 2.8 gram with weight away from outer rim
- 4% of total gravitational energy

Older heavy wheels:
- 3.6 gram with more weight toward outer rim.
- 6% of total gravitational energy

Avoid heavy wheels.
Using older, heavier wheels, the plastic wheels are about 10% of the weight of the car, and hold about 6% of the total energy of the car in rotation energy (doesn't win races) instead of forward motion (does win races).

The rotational energy in the wheels is small compared to the forward energy of the car, but still this energy can mean win or lose.
- The four wheels are only 4 x 2.8 = 11.4 grams (0.4 ounces) total, and the car is close to 140 grams (5 ounces).

Truck tires are big and heavy and harder to turn, but these heavier knobbier tires support heavy loads and grab in mud:
It is actually harder to turn the steering wheel when use double tread heavier tires, because there is more energy in spinning tires, in the form of angular momentum.

Heavy knobby tires for traction in mud and dirt

Trick 5 — Lighter Wheels

The newer 2.8 gram wheels are much better than the older 3.6 gram wheels. You can still drill away some of the mass to reduce rotational energy. Drilling farther out is better because this mass farther out dominates the rotational energy.

Wasted Spinning Energy of Wheels

"Huge wheels are a bad idea if I want to go fast."

We want light wheels, as understood just by looking at round things rolling down hill. The spinning wheels steal some of the limited gravitational energy into spin, and the car goes just a slight, wee-bit slower.

Question: Which goes down slope faster, the ring or the skinny rod?
Answer: Skinny rod beats large ring

Fast (small wheels) **Slow (big wheels)**

Large ring, with lots of rotational energy

Skinny rod, avoiding rotational energy

Slower / Faster

Rings have more Moment of Inertia, even with same mass.

Question: Which goes down slope faster, the Pinewood Derby car with wheels or an ideal zero friction sliding car?
Answer: The block on ideal ice with no friction, but that does not exist.

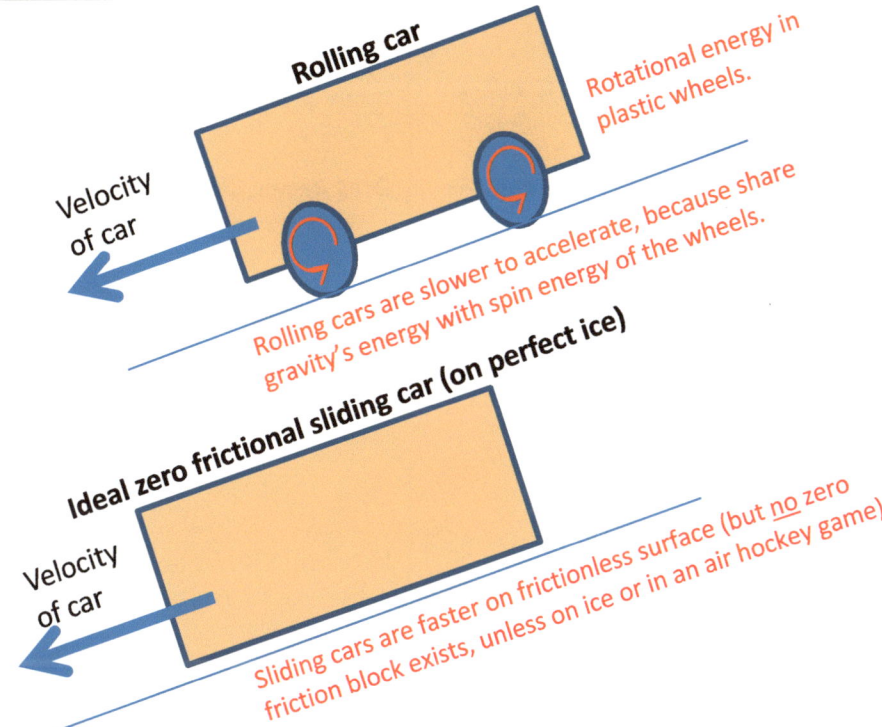

Rolling car — Rotational energy in plastic wheels.
Velocity of car
Rolling cars are slower to accelerate, because share gravity's energy with spin energy of the wheels.

Ideal zero frictional sliding car (on perfect ice)
Velocity of car
Sliding cars are faster on frictionless surface (but no zero friction block exists, unless on ice or in an air hockey game)

Wheels can be different: Thinner wheels with less mass around the perimeter have less rotational energy, so most of the starting gravitation energy goes into forward motion instead of spin.
The object with less rotational energy—the thin rod—goes faster. Most of the gravity energy is the forward energy.

Added weights around perimeter: more rotational energy
Science Museum, Concord, NH

Rotation energy demonstration: Disk with heavier weights around the rim takes longer to go down, because energy from gravity is shared with more rotational energy instead of just forward energy.

Wheels are necessary but steal energy: An ideal block with perfect sliding, no friction, would beat a block with wheels, because the wheels on the block steal some of the precious gravitational energy, as spin of the wheels.
The car with wheels will have same total energy, but the energy is divided up to forward motion and wheel spinning, so the car naturally rolls down hill slower than an ideal block with no friction.

Wheels are great. Cars roll on wheels. Thinner, or lighter, or less diameter wheels will have less rotational energy, and roll down a ramp faster.

Examples of Rotation Energy of Wheels

"My wheels feel so light, unlike Monster trucks."

Pinewood Derby cars have a typical ratio of energy in wheels compared to energy in forward motion compared to people cars. Motorcycles and Monster Trucks, with big fat tires, are the exception.

Pinewood Derby: 4% rotation energy
- Even with light plastic wheels, the car still shares forward motion energy with energy of spinning wheels.
- This 4% of energy is not going into forward motion, and the car does not go as fast.

What's up with big tires?
For people cars, big tires make the car slightly slower to accelerate, because part of the energy needs to go into the spin of tires.

Once you get going, the tire size should not matter for how fast you can go because the energy that is already in the tire stays in the tire.

Larger tires do take more force to start spinning, due to the larger rotational energy in the tire (angular momentum). Stop and go driving is less efficient with large tires.

Truck : 3% rotation energy
- Trucks are not about acceleration, and trucks are so heavy that even heavy ties are just a small part of the weight. Trucks are about grip on snow, mud, and rocks, and high clearance, and hauling heavy loads. So big knobby heavy tires are a good thing.

Truck

Sedan : 3% rotation energy
- Sedans are about comfort, family, and reasonably high clearance in the snow. Hence, moderately sized wheels are good.

Sedan

Drag Racing Car : 3% rotation energy
- Large rear tires are used for more surface area to grab the ground, and support the weight of the engine.
- To accelerate faster, a drag car should have smaller rear tires, because the engine needs to accelerate the car in a quick dash. But designers must have realized fat big tires are necessary to support the engine weight, and grab the ground. Designers do use small thin tires up front, where there is no weight.

Drag race: Large rear tires are used for more surface area to grab cracks in the ground.

Indy car : 5% rotation energy
- The tires on Indy cars need to hold up for 100s of laps. Higher pressure thin tires may need to get replaced too much, losing time in the race.

Indy car

Motorcycle : 10% rotation energy
- Heavy wheels on motorcycles are necessary for balance due to imbalance for two wheels without spin. Fortunately, the engine already is so powerful compared to the motorcycle weight, so heavy wheels are not a disadvantage.

Motorcycle: Spinning moderate size tires help with balance.

Monster Trucks : 10% rotation energy
- Acceleration is not the point: jumping and landing on springy tires is the point.

Monster truck

Rocket car: Rocket Cars can use smaller rear tires because the jet exhaust is pushing the car, not the drive-shaft to the wheels: hence, the tires don't need tire surface area to grab the ground, and there is less rotational energy and the car only puts energy in forward motion.

Chopper motorcycle: Huge tires, but huge power as well, so get comfort and grip.

3% to 5% of a typical car's total energy is in rotational energy. Motorcycles have more rotational energy, because they need the balance.

5.6 Trick 6: Use Exact Maximum Weight of 5 ounces, No Less

"I'm going to ram through that air with all my weight."

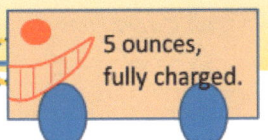

More total weight is good to fight air drag and reduce impact of spin energy of wheels: A 5 ounce car versus a 4 ounces car should have a ½ car length advantage:

- Put some lead weight in the back until the scale just reaches 5 ounces. Use thin weights to keep the car slim for lower air drag. The scale and weights can be purchased at hobby shops.
- Lead weights are easy to use. Lead is a high density, much higher than wood, so small amounts of lead will recover the 5 ounce weight. You could use any dense weight, like copper coins, but lead is also soft and easy to snap to different sizes to trim the car weight to the exact 5 ounces.
- Remember that the center of mass still should be about 1 inch in front of the rear wheels, not behind the rear wheels, or the car will do a wheelie.

The bare block does not get to the full weight allowed. So add some weight.

Two cars that have lead weight in back, to reach 5 ounces and to move the center of mass to the back. Gravitational energy is maximized.

Whatever scale you have, make the weight 5 ounces. Choose a scale that is sensitive at 5 ounces, not one that goes up to 100 ounces.

5 ounce scale: You'll need a scale, where the counter-balance is 25 quarters or 5 ounces.

Bare block: 4 ounces (1.1 Newtons, or 0.11 kg-force)
- Still can add 1 ounce weight to a basic block.

Weighted car: 5 ounces (1.4 Newtons, or 0.14 kg-force)
- Used lead weights

A heavier car does not get pulled (accelerated) by gravity any faster than a lighter car. All objects fall at the same rate 'g' in a vacuum.

A heavier car will have more force from gravity, but also more mass to accelerate. So acceleration stays at the same value 'g', independent of mass.

Why strive to get to maximum weight? A heavier car can push through air drag with more weight, so air friction does not slow the car down as much. Also, a heavier car body makes the wheel spin energy just a smaller part of the total energy by having most mass of the car be the car body, not the wheels. You'll be a ½ car length faster at 5 ounces than a car at 4 ounces, due to these air drag and wheel spin reasons.

More weight, for same size car, means that the car has more energy and can plow though the same air drag easier. In addition, the wheel rotational energy will be a smaller fraction of the total forward energy.

Heavy Cars Plow Through Air

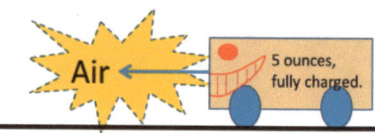

"I'm like a rolling battering ram, with all my weight in my ram."

Let's list the reasons that heavier cars go faster.

Heavy cars plow through air drag:
Air drag force will slow a lighter car more than a heavier car.

Heavy cars put more gravity energy into the forward movement of block, not the spinning wheels:
Wheel spin energy is still stealing energy from forward speed. If the block is lighter, wheel spin energy steals a larger fraction of the energy.

Out of this world example: dropping stuff on the Moon without air getting in the way.

If your Pinewood Derby car is rolled in a vacuum, like on the Moon, then weight would not matter so much: no air drag, so any weight would plow on. Only the rotational energy in wheels would be lost.

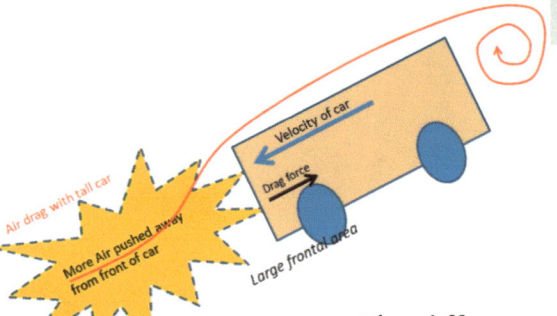

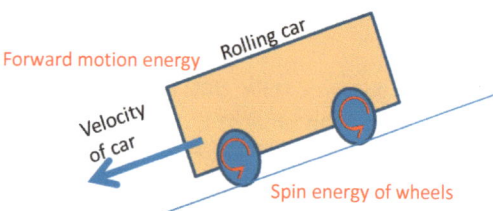

The difference is ½ a car length, between a 4 ounce car and a 5 ounce car.

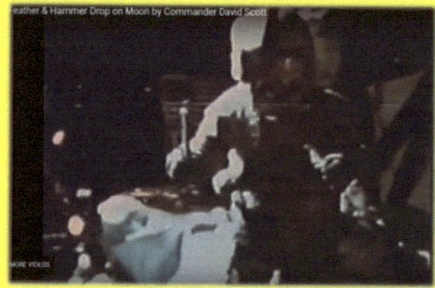

Feather and hammer ready to drop on Moon in a vacuum (before high resolution videos)

Reasons to give the car the maximum weight allowed:

- **Air drag:** For the same shape car, air drag is the same whether the car is heavy or light. The same air drag force will slow a lighter car more than a heavier car.

- **Wheel spin energy:** Energy in spinning the wheels does not reduce when the car has less weight in the body. With lighter cars, there is less percentage energy going into forward velocity.

Air drag and wheel energy stay the same and are not dependent on mass. Air drag slows a lighter car down faster. A lighter body, combined with fixed mass wheels, means the car does not get up to as high a speed. Monorail drag and axle drag are proportional to the mass, so these other drag forces are not impacted by using less weight.

Hammer   Feather

Air drag and gravity experiment on the Moon:
Hammer and feather hit the ground at same time on the Moon without air drag, during Apollo 15 mission. On Earth in air, the hammer would plow through the air with no trouble and the feather would flutter down.

https://www.youtube.com/watch?v=5C5_dOEyAfk

Trick 6 — Use Maximum Weight

A 5 ounce car gets slowed down less by air drag and wheel energy, for an ideal car.

5.7 Trick 7: High Bumper and Paint Job Minor Tweaks

Here are ways to get both a head start with a raised bumper and lower air drag with a smooth car, although these tweaks are not as important as going straight, grease, and car shape to get low air drag.

"I'm like a human sprinter getting the jump on the starting gun."

Higher front bumper to get a head start:

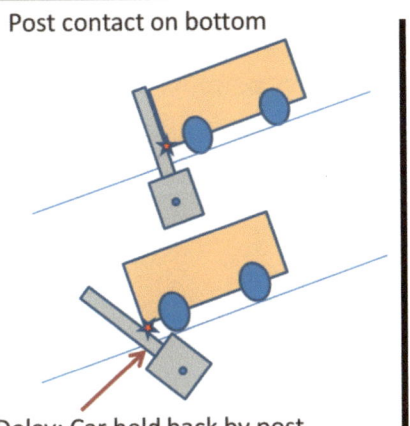
Post contact on bottom
Delay: Car held back by post for about 0.05 seconds.

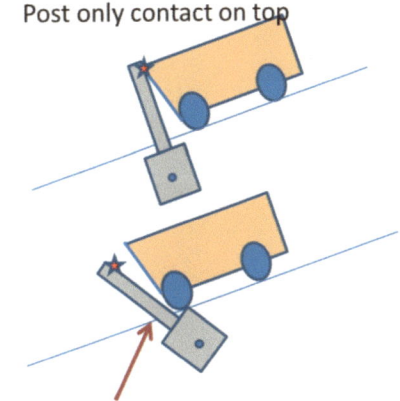
Post only contact on top
No Delay: Free rolling

Have a high bumper in the front, so that the car starts a little before the other cars, during the short time it takes the starting post to drop down.
- However, air drag raises it's ugly head. Now the front area is larger and the air drag is larger, so a raised bumper is probably not worth the trouble if the contact point is any higher than the car.
- Also, if 'Quick Draw' Scout Master is running the race where you hear a snap or shock wave as he snaps the posts down, then this high bumper is not helping.

Car with high bumper

Glossy paint job to lower air drag:

Supposedly air plane speed records were made, at speeds much faster than PD speeds, by sanding down the screws around the fuselage. So maybe, just maybe, smooth surfaces work for Pinewood Derby cars too.

Car with smooth paint job

Will dimpling the car surface reduce air drag, like golf balls?
Dimples hold the air stream against the ball longer, so less turbulence behind. There is less of a low pressure behind the ball, and so less drag.

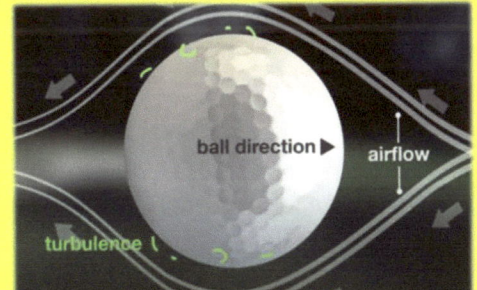

If had turbulence behind ball, then have less rear pressure and have rear air drag.

Ideal air flow past dimpled golf ball, with less turbulence on the trailing end of the ball.

These tweaks are debatable, and maybe more Pinewood Derby 'lore', but you can try them.

Trick 7 ✴ Higher Front bumper

5.8 Trick 8: Dumb Luck and Foolishness, the Human Factor

After all the work making sure the car goes straight and the wheel axles are well greased up, let's not forget that the car is a toy that gets 'practice runs' between getting built and getting handed over immediately after the check in. Accidents occur, just like a teenager getting their first car after squeaking by on their driver's test.

Don't forget Pure Dumb Luck
- Aligning the car straight at beginning (first few feet will rub against monorail)
- Getting on quickest lane (some lanes have rougher monorails and seams), which is why cars are switched between lanes between races.
- Not dropping car before the race (crashes can mis-align the wheels)
- Make sure your scout doesn't try to bang up the car before the races, during practice runs.

"A sign that Dad is taking the Pinewood Derby too seriously:
- Dad threatens to break the scout's fingers if he touches the car after the wheels are on."

"Oops, sorry!"

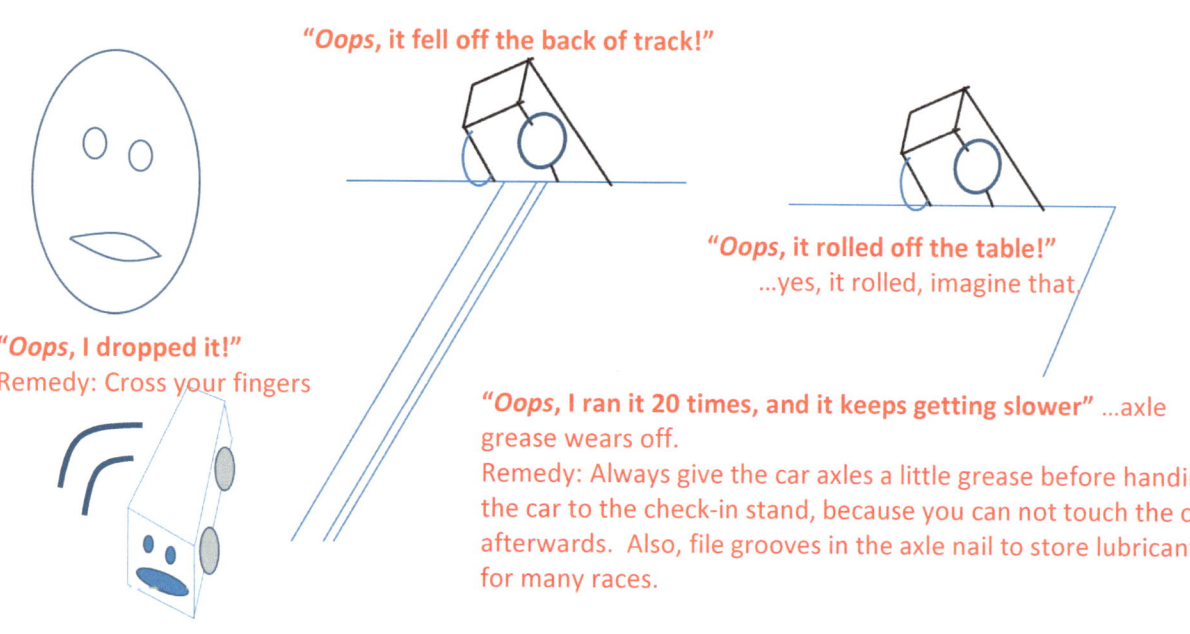

"*Oops*, it fell off the back of track!"

"*Oops*, it rolled off the table!"
…yes, it rolled, imagine that.

"*Oops*, I dropped it!"
Remedy: Cross your fingers

"*Oops*, I ran it 20 times, and it keeps getting slower" …axle grease wears off.
Remedy: Always give the car axles a little grease before handing the car to the check-in stand, because you can not touch the car afterwards. Also, file grooves in the axle nail to store lubricant for many races.

It is fair to warn your scout that racing the car many times before the official races can cause accidents, mis-aligned wheels, and that the scout has the opportunity to race after the official races if they want to. The scouts probably won't listen, and they'll just want to run races themselves of their car against their buddy's creations.

These accidents are going to happen, with horseplay and the friendly challenge. Be the hero, and remember to bring a few tools to the races to fix what you can. Also, if you are the parent, your scout's Pinewood Derby car is not your car and your kid can judge for themselves how much they care if the car rolls off the table.

Real car accidents and fender benders

Oops, I hit a pothole.

Oops, I ran over the curb.

Oops, I rolled down the hill.

Trick 8 — Don't crash before races

Dumb Luck and Foolishness, the Human Factor

Here are the main two things to do – go straight and get rid of nail axle friction. Both are at risk due to the human factor. A crash can bend a nail / axle and cause monorail rubbing and drag because the car now turns. Too many runs causes the axle lubricant to wear away, so the car gets more axle drag.

"A sign that Dad is taking the Pinewood Derby too seriously:
- Dad threatens to break the scout's fingers if he touches the car after the wheels are on."

"Oops, sorry!"

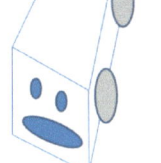

Glued nails

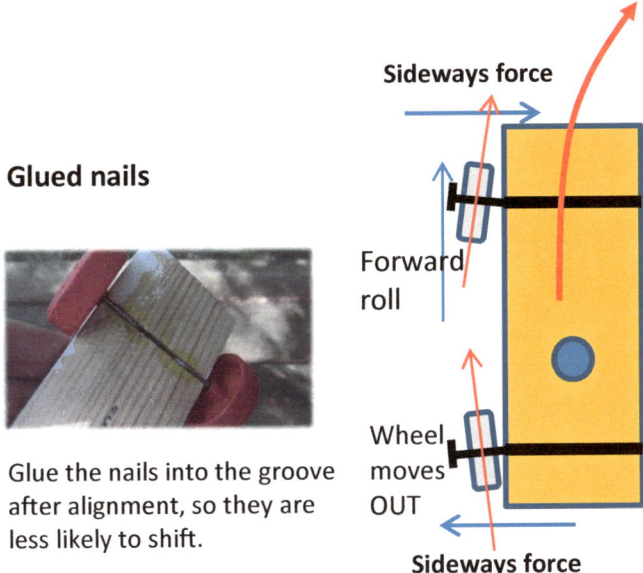

Glue the nails into the groove after alignment, so they are less likely to shift.

Squeeze graphite on and spin wheel

Wear and tear:
Over many races, the nails can move inside the grooves and mis-align the wheels.

Loss of axle grease:
We make a reservoir of graphite powder or axle grease, but the reservoir can run low after 10 fun runs before the official races.

More trial runs means that the lubricant wears away and the car slows down.

Chapter 6: Fastest Track Shapes

Who decided on the track shape? Well, basic physics did, to get the most speed right off the bat. ...and the fact that straight flat sections of track are easiest to build. We want to coast at the faster speed for most of the track.

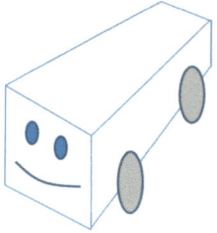

Build up speed quickly by doing an almost vertical fall.

Coast at top speed over most of the track.

The idea of a steep fall at the beginning to pick up speed is not new. To finish the race a quickly as possible, we want the cars to get up to speed as quickly as possible. A very steep drop at the beginning might add a little extra travel distance, but those tracks are quicker because the cars get up to speed as quickly as possible.

The first cliff hanger section can't be too steep because the PD car needs to be able to handle the bend at the bottom to get onto the straight away.

The final speed under gravity to get down from any height, in any path possible, is just a function of the height change, not the path. But if the path is very long then the travel time is very long. Also, a longer path just means that there is more length for all the drag forces to slow the car down.

There are examples in real applications that a steep ramp or drop is used to quickly get to speed. Planes were dropped from blimps in the early 1900s, and picked up speed to get air flow and lift. Ski jumpers start with a steep ramp. Possibly a aircraft carrier could use a ramp to get airplanes to a faster speed, instead of a catapult.

A steeper start on ramp would have faster race times, but the wheels might pop off at the bottom of ramp at the bend to straight-away. Also, for racing, any track shape is fine, because all cars are equally competing on the same shape.

Different Track Shapes

Who decided on the track shape? Well, basic physics did, to get the most speed right off the bat. ...and the fact that straight flat sections of track are easiest to build. Without any losses, all these tracks below will have the same speed at the end, but different tracks will immediately get to that speed and others won't.
The slowest time happens with just a gradual ramp. There is a delay getting up to speed. The fastest time happens when the car immediately drops the car below the floor. That requires a raised track.

"The steep ramp at start will build up speed right away and make the race exciting.
The flat straight-away is a great way to challenge straight driving cars."

What track shape is the fastest?

Slowest time, even though the shortest path
A gentle slope takes longer for gravity to get the car up to speed.

Pinewood Derby track shape

Immediate cliffhanger kick-start for car, to get up to speed immediately.

Fastest time, even though the longest path
Want to get highest speed as quickly as possible: make gravity work quickly and then coast at a fast speed.

What slope track is the fastest?

Different slopes to race against

Steeper slope gets to bottom first
Even though a higher ramp has more distance to travel, the steeper car accelerates faster and the car gets to the bottom of the ramp faster.

Track at Regional competition
The flat straight-away is the best compromise with practical builds of a track and the initial steep ramp to get up to speed.

Like Pinewood Derby track.
The red brachistochrone (inverted cycloid) curve is the curve of fastest descent between two points
Steeper start gets the ball to the end sooner, because the ball can coast faster sooner, even though the track length is actually a little longer.

Looped track (not official)
A loop in the track requires a very fast car to have the speed to stick to the track even when upside-down. The starting height is well above the typical 4 feet.

The shortest distance with a gradual ramp is actually the slowest, because the car doesn't build up speed quick enough. We want the sharp drop right at the beginning to get up to speed right away.

Different Ways to Use Gravity

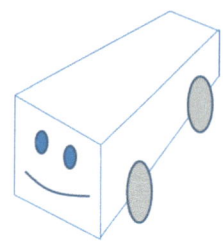

Who decided on the track shape? Well, basic physics did, to get the most speed right off the bat.

The curved crazy street in San Francisco requires a lot of braking, so the speed is limited to follow the bends.

An airplane can drop almost vertically, picking up much more speed than the engines could provide. The pilot must have quite an adrenalin rush. It is nice to be like a bird in the air.

San Francisco Lombard street: Most crooked street in the world.

Carpet slides at carnival

Fighter planes getting extra speed by nose diving

The final speed under gravity to get down from any height, in any path possible, is just a function of the height change, not the path. But if the path is very long then the travel time is very long.

The crooked San Francisco Lombard street has a lot more distance to travel to get down to the bottom, so the race time to get to the bottom will be longer. Again, the final speed by coasting, ignoring braking around turns and obeying the speed limit, will be the same with or without the turns.

Carpet ride wavy slide at carnivals also has a longer distance to travel to get to the bottom. The final speed will be the same, with or without the waves, but the travel time will be longer.

Dogfights in WW2 also took advantage of the extra speed gained by starting higher than the enemy airplane.

Roads for people don't want the steep drop. Maybe skiers want a starting steep drop, if they have a crazy streak. Airplanes, like fighter jets, can easily use a steep drop.

Chapter 7: Odd Cars, like Cars for Fun

Free time on the track: It is always fun to race your cars before the real races begin. Feel free to try anything, although only cars that follow the rules (weight, length, height) can compete in the official race.

"Good times when going with your heart, instead of speed."

Flap to engage music

Top music-card flaps open from the air flow pushing the flap, and the music begins, just like a birthday card.

Here are the electronics from inside of music card.

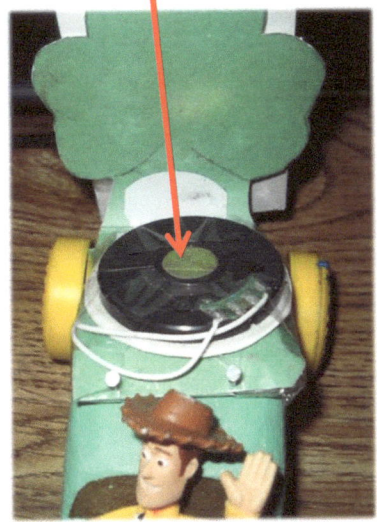

Electronics scavenged from music card

Odd cars ready to race, for trial by fire

Try any design, even if the design can not officially compete due to the rules.

The green music-card car might even be legit, because the flap is attached. The falling weight white car is not legit, because there are loose parts.

Real world fun cars

Sometimes it's the classy beach cruiser fun more than the speed, like tricked out cars going along some trendy beach drive.

Parade Cars

Upside-down VW Beetle Candy bunny

Have fun with music, cartoon characters and paint, and experiment

Enjoy the time at the races Listen to the music card

75

Race of the Odd-Balls and Misfits

Odd-ball, non official car races: Here is the optimized slim car, 5 ounces, versus an odd-ball Drop Weight car and the Music Flap car. The optimized slim car wins, although only by 1 car length.

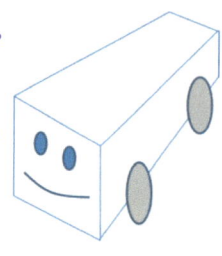
"Hey, the lower profile car will be faster, but everyone loved the music car with a flap."

Practice runs without rules.

Weights are falling out of ladder.
Loose parts, not legal
legal   legal

Music car is opening the flap from air drag

Odd cars rolling down ramp, near neck and neck until air drag forces taller cars behind.

Regular slim design will win, because of lowest profile and air drag.

Out of the gate, these regular car versus odd-ball cars are neck and neck, and they stay close even at the finish line. So the air drag of the odd-ball 'flipped over flap' car and the 'weight holder' car are not that big a deal, although they still lose.

These cars, good or bad, do not have consistent speeds over many runs, because nail axles can be bent between races and graphite lubricant can wear out after racing 10 times.

The low profile car wins, as expected due to lower air drag.

If you want music, include a music card flap. If you want not 'street legal' physics experiments, then do that. Or add a rocket, a propeller, or wings

Chapter 8: High Pressure CO2 Powered Rocket Cars, on Straight-Away

You can add rocket power to a Pinewood Derby, just for kicks, when you graduate from Cub Scouts into the Boy Scouts. Now the competition is called CO2 rocket cars.
The races are done on flat ground, and don't depend on gravity. The car is guided by a guide rope the entire way along the floor.

Hole for CO2 cylinder, typically used for pellet guns.

CO2 car with only one cartridge.

Eye screws on bottom for guide rope.

"These are Boy Scout races, with rocket power. Older scouts graduate to powered compressed gas CO2 cars on straightaways (flat floors), no gravity.
This exhaust thrust makes it a rocket."

High pressure gas CO2 powered cars use exhaust gas like a rocket for thrust. Because there is rocket exhaust and thrust, these cars do not use gravity, and travel on a flat ground along a guide rope.
For these rocket powered cars we want the least weight, so the CO2 exhaust will push on less mass. This is not a gravity car. The rules and physics are different, and horsepower to weight ratio now matters.

* From internet

Single CO2 cartridge or cylinder design

Ways to make a quick hole in the cylinder cap to start race?
- Hit puncturing pins with a hammer
- A bumper resting against the starting gate releases a mouse trap which punctures the CO2 cartridges.
- If on the sloped Pinewood Derby track, you could make a parachute flap contraption which releases a mouse trap, to trigger the puncturing pins to pierce CO2 cartridge.

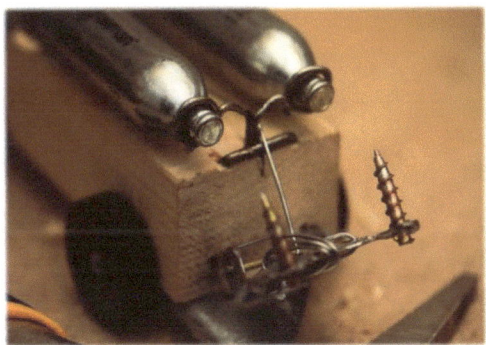

Double barrel thrust with two cylinders

Is a fan powered by CO2 faster?
Probably, like a propeller on an airplane is more efficient at lower speeds compared to a jet engine (turbo jet).

The older Boy Scout cars can go faster than gravity, using CO2 rocket power!

High Pressure CO2 Gas Rocket Force

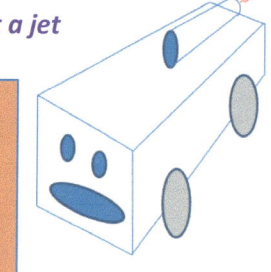
"Yeah, I got a jet pack on!"

These double barrel CO2 rockets are ready to fire, for this Boy Scout activity! Just hammer them open at the same time, and get double the force and acceleration.

Unlike the gravity powered PD cars, the mass of the car has a big impact compared to the fixed thrust of the CO2 exhaust. We are talking about horsepower to weight ratio, just like any powered car. The thrust is not from gravity but from a rocket engine, so less mass means more acceleration and speed. Gravity force is proportional to mass, and rocket thrust is not. Rocket thrust is proportional to the exhaust rate.

Gas pressure CO2 canisters ready for puncture

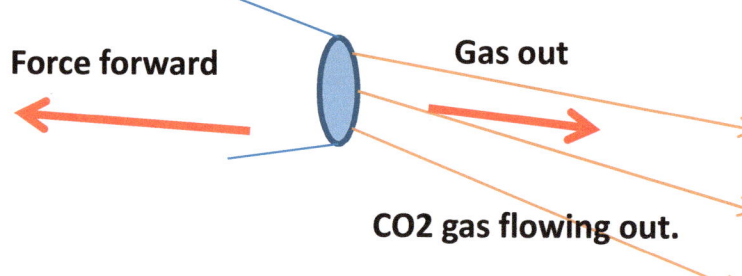

Force forward — **Gas out** — CO2 gas flowing out.

The CO2 gas should accelerate the car just like a rocket engine, even on level ground, with about 2 pound-force. Assuming the gas does not empty out, a larger diameter hole will cause the car to go faster right from the beginning.

The CO2 rocket force or thrust of 2 pound-force is much larger than the gravity force on a Pinewood derby car. And the CO2 rocket force keeps going until the gas runs out.

The main use of CO2 cartridges is for pellet guns:

Pellet gun: Typical use for CO2 cartridge is a high pressure push for a pellet gun.

More typical chemical rockets, not CO2:

Fireworks

Estes toy rocket

Space rocket: the ultimate rocket with over 90% fuel.

High Gas Pressure Rocket Inspirations

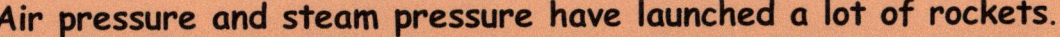
"Yeah, I got a jet pack on!"

Air pressure and steam pressure have launched a lot of rockets.
High pressure air will always work, although the energy is low. The exhaust will always shoot out after the valve is opened, no ignition is required.
Chemical fuel has more risk. Chemical fuel needs to be ignited, or is acidic, or does not store well. So for applications that only need low thrust, compressed air is a good choice.

- Steam rocket
- Space SAFER mobility
- Satellite station keeping air blasts

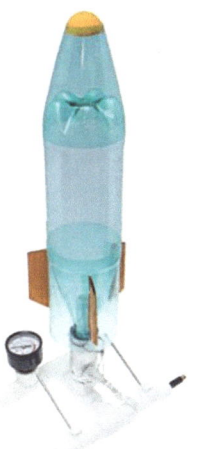
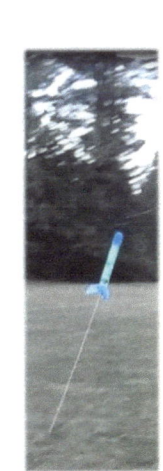

Spacewalk: SAFER jetpack in space, no tether.

SAFER jetpack which goes over the life support suite.

Test firing of steam rocket

Rocket Launcher for Water and Soda Bottles. H- Base/U-Trigger/Pressure-Gauge
Brand: WATER ROCKETS CLUB

Compressed air is shot out of 24 gaseous-nitrogen thrusters to move astronauts around the ISS in an emergency separation.

Here is the spacewalk in orbit, using the SAFER jetpack, which can blast compressed nitrogen to steer the astronaut back to the space station.

This air exhaust rocket engine is real, and it is used in space to help astronauts get back to the Space Station, or to keep satellites oriented in the right direction.

Chapter 9: No Rules 'Outlaw' Pinewood Derby Cars

Rocket power? Air power? Back-fire recoil? Swinging ball power? See the next few pages to see how each car below performed in the quest to out-race a regular Pinewood Derby car.

"Hey, this isn't fair …they've got rockets, propellers, back-fire 'bullets', and moving parts!"

In PD races, the standard ways to make cars go faster are not allowed, like engines.

The PD rules almost invites, or begs for, a 'no rules' category. People are just thinking, why not use a bearing? A rocket engine? A propeller?

Psychologists get people to eat sometimes by denying the food. In old England, the government outlawed stealing potatoes from the King's field, so then the peasants desired and stole the potatoes.   Well, people wanting to have fun with PD cars will try other things.

You might be tempted to strap on a model rocket engine onto the back of the PD car. With rocket thrust, we don't need to depend on a downward ramp and gravity. Also, these model rocket engines produce a lot more thrust than the weight of the PD car.

Well, there sometimes are no-rules races, typically entered with cars by the adults in Pinewood Derby. This chapter experiments with a few cases.   A regular pinewood derby car is also raced down a ramp against a no-rules modified car, and there is a video of the race. The regular PD car is the 'control' car. With a side by side race, it is obvious which car is faster, or not faster if the experiment is a failure.

The rocket engine wins hands down. The rocket car now has gravity and rocket thrust going for it. The gravity car barely moves by the time the rocket car has reached the bottom of the ramp. The rocket engine has thrust by ejecting stuff backwards.

The propeller mechanism was scavenged from an air powered propeller airplane, and that still works better than just gravity.   Propellers work by pushing air backwards.

The throwback car involves a precariously balanced weight on a mouse trap, that is released as the car starts to move underneath. This throwback car works by pushing a heavy weight backwards, like recoil on a gun.

Some of these ideas can be dangerous and need a keep out zone, because exhaust or weights are thrown back. Don't have anyone stand behind the throwback car, or the rocket car.

There are many powered Pinewood Derby car ideas using rockets, fans, rubber bands, for the fun and frustration outlet of all.

No Rules 'Outlaw' Pinewood Derby Cars

Rocket power? Air power? Throw-back recoil power? See the next few pages to see how each car below performed in the quest to out-race a regular Pinewood Derby car.

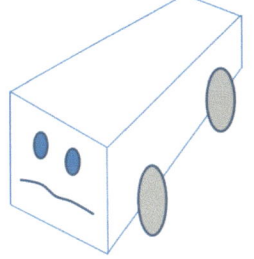

"Hey, this isn't fair ...they've got rockets, propellers, 'bullets', and moving parts!"

1: Rocket (chemical) powered

Rocket car and gravity car ready and behind the starting block.

Works Great: On a 5/16 lb car (5 ounce), the rocket should cause 5 times more acceleration than gravity down the ramp, with a thrust of the rocket engine of 1.3 lb force.

2: High Pressure air powered fan

Air powered propeller on straight-away

Works good: Top of toy air-powered airplane taped to Pinewood Derby car.

Other Outlaw ideas

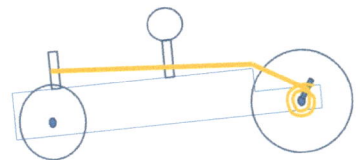

Rubber band car

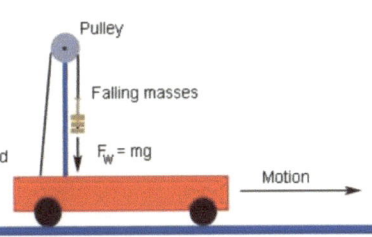

Weight on string car

3: Back-fire mouse-trap car

Standard gravity-car comparison during race

Race a back-fire car against a standard car. Back-fire car has flying weight on top of mouse trap.

Even more ideas:
- Electric motor powered car, with gears on shaft, like remote-controlled toy cars.
- Autogyros use air drag from forward motion to spin helicopter blades and get lift.
- Some external pods on airplanes use jet turbines to generate electricity (to power electronics), powered from blowing air like a windmill.

Electric motor and propeller car

There are many powered Pinewood Derby car ideas using rockets, fans, rubber bands, for the fun and frustration outlet of all.

Heartbreak: Rocket Car Race-off Against Pinewood Car

Whoa ...no competition!   Rockets are in a whole different league.
Very fast, with rocket engine ...gravity, eat my dust. The rocket car went 8 feet in the time Gravity car went 2 inches.
This is one reason there are restrictions and rules, against rocket engines, electric motors, and rubber bands, to level the competition.

"Eat my dust, gravity car!"

1 Ignite rocket, with electric current through an igniter

2 Release gate, that ensures that both cars start at same time

Rocket car behind the starting block.

3 Out of gate: Observe completely different league of racing

4 Gravity powered car moves an inch so far, while the rocket car is nearly down the whole ramp.

"You do know that rockets are disqualified, right?"

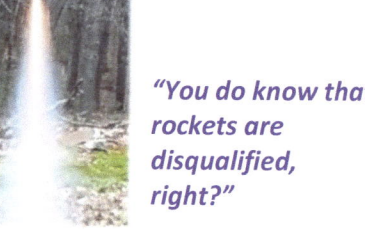
Static test of toy model rocket exhaust gas

Inspiration: a dangerous looking rocket on a car.

"Did the race start yet?" asks the gravity powered car.
...Gravity car gets heartbreak

Ratio of distances 'd' at an instant in time gives the ratio of the forces:

$$\frac{F_{rocket}}{F_{gravity}} = \frac{d_{rocket}}{d_{gravity}} > 10$$

Chemical rockets provide an exciting route against gravity, especially on a gently sloped ramp. The force from the rocket does not care about gravity or the ramp.

Chemical Rocket Inspirations

Whoa ...no competition!   Rockets are in a whole different league. Very fast, with rocket engine ...gravity, eat my dust.
Again, this is one reason there are restrictions and rules, against rocket engines, electric motors, and rubber bands, to level the competition.

"Eat my dust, gravity car!"

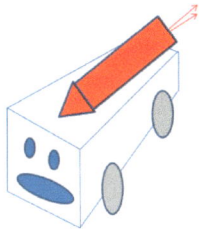

- Space rockets
- Jet engines

Rocket on bicycle:
Avoid the pedaling, with no power directly at the wheels.

American Apollo space rocket.

Static Test of Thrust for Estes toy rocket engine:
This model rocket engine has about 1 pound thrust, pushing up a 0.1 pound rocket

Rocket sleds:
The fox gets the built up speed, but he can't turn fast enough to catch the slower roadrunner.
This large turning radius is a real effect when going fast. Slower rabbits escape charging foxes, by zig-zagging around.

Chemical rockets are loud and fast and have a thrust much larger than the force of gravity down the ramp. Chemical rocket engines are designed to lift toy rockets vertically against gravity.

Pressurized-air Plane Propeller Car Race-off and Propeller Bragging Rights (Success)

Here's a head to head race of a propeller-assist car and a gravity-only Pinewood Derby car. Of course, with some good propeller thrust assistance, the air power wins the day.

1 Pump high pressure air into plane bottle, and flick propeller to start the spin.

Start propeller by flicking it

2

Release block gate

"Add a powered propeller to your Pinewood car, and win …of course, you are disqualified…"

High pressure air (~50 psi)

Combined PD and scavenged body of air power airplane model: Propeller, spun with air pressure into a piston engine

3

Propeller car has gravity and propeller thrust

4 Plane car picks up speed faster than gravity.

Propeller car keeps pulling ahead

5

Propeller stay spinning the entire race

6 About twice as fast out of starting gate.

Propeller pulls ahead, and will keep thrust on the straight-away.

Air Hogs pump plane, with pump

Inspiration: Getting around in the Florida Everglades, or propeller backpacks.

The Propeller car works. The propeller adds a little bit extra force, on top of the force of gravity. The gravity force is still larger than this propeller force.
At the bottom straight-away, of course, the gravity force is zero and the propeller still has force.

Propellers and Pressurized-air Inspirations

Normally a propeller is powered by an engine, like the Bayou Swamp Airboat. But the propeller can also be powered by compressed air.

"Add a powered propeller to your Pinewood car, and win ...of course, you are disqualified..."

Bayou Swamp Airboat

The propeller car gets the benefit of both gravity and the thrust of the propeller. We are talking about the same level of thrust as a rocket burning the fuel really fast, but the propeller thrust is clear.

Air pressure is used in tires, in machine shops for tools, for scuba diving. Air pressure can also keep an engine running, similar to a steam engine.

- Compressed air cars and motorcycles
- Steam locomotives

Experimental uses of compressed air for transportation

Power Cars: Compressed air is used to power some cars in India

Power Scooters: Compressed air can power a scooter

Pressurized air Rocket car

Steam engines

Tools in car garage and machine shops: Pneumatic tools spin using compressed air

Air Hogs pump plane, with pump. The bottle body with air powered piston was used for the propeller.

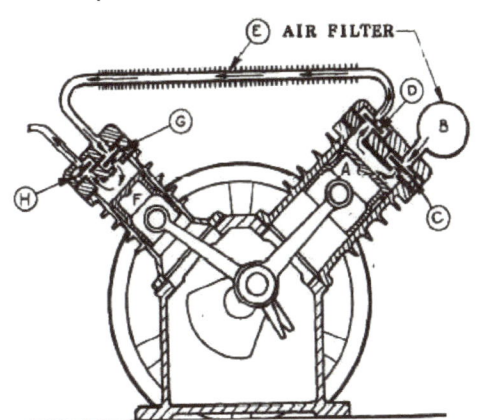

Pistons powered by air pressure instead of explosions

Rotating engines can be powered by high pressure air.

85

Back Fire Car: Throwing Car Race-off and High-kicking Rodeo Moves

"Throwing lead out the back using a mouse trap works, if you can keep the car on the track."

Another version of a rocket is to throw stuff out the back, just without the flames. That mousetrap snaps down, flinging the weight backward. The car flips up, and we're just lucky the flying weight did some good. Whoa! Get ready for a show!
Remember, the rules say no flying parts, so we're disqualified at the start. And, of course, don't stand behind the car.

1

Motion starts when gate released

2

Mousetrap triggered by tilted 'cheese' weight

Back-fire car has flying weight on top of mouse trap.

3

Flip in air
Mousetrap swings down and throws weights, car swings up

4

In air flight, back of car rising up to meet the trap

Throw lead weights using mousetrap. Mousetrap triggered by balancing weight which tips over onto 'cheese' trigger at start.

Car flips up! …just like any mousetrap does when it snaps. The car by chance lands back on the track.

5

Inspiration: Recoil against your shoulder from potato gun
Accidentally land back on track
Car lands back on monorail, by chance, after bucking up

6

Mousetrap car keeps the lead, but only with gravity's help for remaining length of ramp

Lots of energy lost when mousetrap car snaps up …but we want car to ride flat. The car is half snapping up to the mouse trap, instead of just the bars of the mousetrap swinging down.

The Mousetrap car is about 30% faster than the regular gravity-only car, if lucky and the mousetrap car lands back on track after bucking while flinging the weights backward.

Recoil works to give mousetrap car a head start, like a rocket. After recoil, there is no extra pushing.

Back Fire Car Inspirations

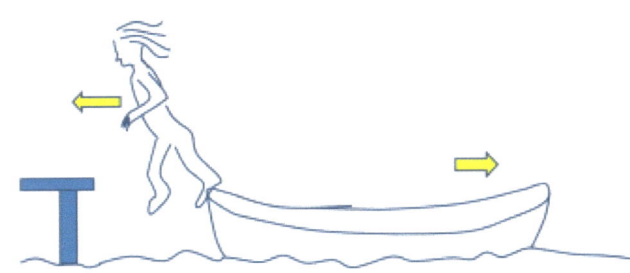

"Throwing lead out the back using a mouse trap works, if you can keep the car on the track."

> **That mousetrap snaps down, flinging the weight backward.**
>
> Many designs throw something back and get a recoil or thrust -- guns, cannons, arrows, rockets, just walking and pushing off the ground.

Guns and Cannon thrust or recoil or 'kick back':

Guns have a rocket force too, called the recoil. A bullet (exhaust mass) is getting shot out the barrel (nozzle exhaust from bottom).

  $Mass_{pellet}$

potato gun

$Force_{back}$ $Force_{pellet}$

'Reaction'   **'Action'**

← The gun and shooter and floor get pushed backward.
← Rocket thrust

→ Pellet gets pushed forward.

→ Exhaust gas

Recoil is a form of rocket thrust:
- Hand gun jerks backward and up when fired, from the kickback, pivoting around shooter's hand.
- For rifles, the shooter holds the rifle butt against their shoulder so gun doesn't have a gap to fly back and 'punch' the shooter.

Falling in water when pushing off a small boat:

<u>Boats</u>: Push off the end of a canoe and see the canoe go the other direction. You are the 'exhaust'.

Pushing off a canoe, by stepping out of the canoe, shows Action and Reaction

It is hard to step out of a canoe, without the canoe moving backward out from under you.
...be ready to get wet.
This backward push is an example of action and reaction, or conservation of momentum.

Anytime something gets pushed away, there is an opposite and equal force backwards.

More 'Other' No Rules Cars, Above and Beyond PD

Try out your own ideas and Frankenstein contraptions. They might not all work, but they sure are fun. Batteries can power fans. High pressure air can power a propeller using air powered piston engines. And, of course, rocket engines release all the energy with a big burn.

"Go crazy!"

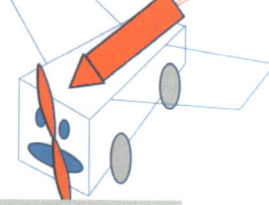

Fan powered by battery

Fan and battery thrust

Fan and battery thrust

Take a computer fan, or use a toy motor and propeller.

Bottle rocket, powered by high pressure air

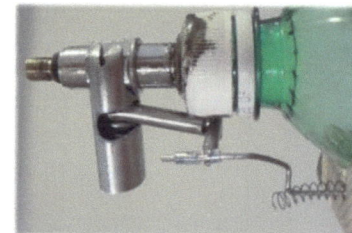

Air exhaust like a rocket

Propeller, powered by air pressure:
Body of toy air-powered airplane taped to top of Pinewood Derby car.

Rocket, from chemical engine

Toy Model Rocket engine, blasting out the back of the PD car

Exhaust of rocket engine in a static test

Thrust of rocket:
The rocket should cause 8 times more acceleration than gravity down the ramp (2.5 ounce-force).
- Use a typical 1.3 lb force model engine (21 ounce-force) on a 5/16 lb car (5 ounce-force)

Electric fans and air bottle rockets help boost the speed of the car. Chemical rocket engines help the most of all.

Use Any Transportation as Inspiration

Here are some sources of inspiration from real 'modes of transportation'.

"Take inspiration from the world!"

Car inspirations:

- **Turbo-fan jet engine**
Who wouldn't want to borrow a jet engine. The car keeps its thrust even at high speeds.

- **Steam engine**
Steam engines have a long and nostalgic history. The American West was populated with steam locomotives.

- **Air rocket**
Air pressure just takes a pump, and the car has power.

- **Chemical fuel rocket**
Which would win: a rocket dragster, or combustion dragster?

- **Air propeller powered boat**
Do you need to get around a swamp with lots of shallow water, roots, and alligators? Then you want the propeller above the water, not below.

- **Mousetrap car or Spring car**
Springs can have a lot of energy, and release it quickly. Springs are used in old windup watches, and in garage doors.

Turbofan engines

Steam engines

Steam punk

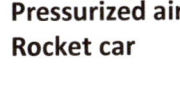
Pressurized air Rocket car

Chemical Rocket car

Bayou Swamp Airboat

The 'Outlaw' race

Spring wound car

Come up with your own 'outlaw' car, to beat a 'street legal' car.

Chapter 10: Summary

The Pinewood Derby car brings up some many concepts. The ideas we've talked about all come up somehow during the competition, such as wood craftmanship, friendly competition, the track shape, driving straight, and some wheel energy and height concepts.

We learn about the craftsmanship that goes into woodworking, and the pride of following through on a design and creating a beautiful car out of an idea. Take your favorite car or cartoon character or fruit and design a car around it. Scouts can use woodworking tools and enamel paints. Parents can help too, or make another car side by side with the scout.

We learn about friendly competition. There's a lot of fun racing against friends at the Pack meeting. Scouts can each talk about their own car idea, and show the other scouts how straight their car rolls.

For the beginning racer, the apparatus of the racing track is also new and exciting. The mechanism of flipping the stopping posts down is new, and the way to stop cars at the bottom using a pillow is new.

The main design goals for the PD cars are to reduce the monorail drag by going straight, to reduce the axle drag by adding graphite powder or some other lubricant, and to reduce air drag by a lower profile and streamlined car.

We learn about making the wheels roll straight. Make sure the wheels don't roll to the side as the car rolls forward. If they do, then bend the nail axle slightly or add a shim.

We learn about shaping the car to reduce air drag. Lower frontal area, or lower profile cars, will have less air drag. A streamlined shape, like a tear drop, will also lower the air drag.

Maybe we also learn some energy concepts, like the energy gained by rolling down the ramp, and the energy lost due to drag mechanisms. There is no engine, and all the energy for forward speed comes from gravity and rolling down the ramp. This is no different than downhill skiing or sledding.

To have a different kind of fun, we can also ignore PD rules and build 'outlaw' cars that use springs, or rocket engines, or fans, or falling weights.

Starting block, for your imagination
Scouts get the official wood block and the rest is up to them.

Pack Races

Six step process to build a winning Pinewood Derby car:
Reduce drag:
1. **Monorail rubbing and drag: Align wheels**
    - Don't crash and bend wheels before races
2. **Nail axle drag:**
    - Graphite on the axles
        - Sand grooves in nail to store the Graphite
    - File flange off the inside nail head.
3. **Air Drag or wind resistance:**
    - Low (thin) profile
    - Wing-like shape

Increase energy slightly:
4. **Center of Mass in back and higher for more gravitational energy:**
    - Weight toward back
    - Remove wood from middle so more weight in back
5. **Use all the 5 ounces allowed to plow through air drag**
6. **Lighten wheels to reduce spin energy, and increase forward energy.**

Starting energy up top: The energy at the bottom can not be more than this.

Finishing energy at bottom: The energy at the bottom can not be more than the starting energy.

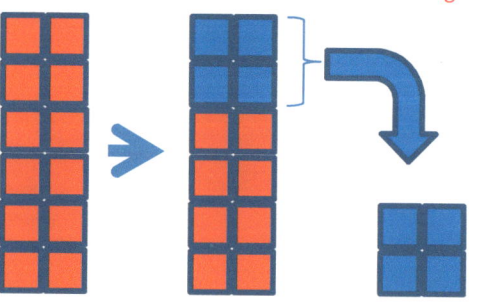

Potential Energy from gravity at top

Kinetic Energy (motion) at bottom if no friction

Energy lost to friction (monorail rubbing and drag, air drag, axle drag)

Priority #1: Make the Car Roll Straight

When the car is constantly trying to turn into the monorail, that is the same as riding the brakes of a bicycle or car all the way down and along the straight away. Don't let that happen.

The drag against the monorail is by far the biggest source of drag and slowing down.

Step 1: Tune individual wheels
Roll the car forward while looking down on it, and see if any of the wheels gets tugged either toward the car body or away from the car body. Those are the wheel axles that need attention. That attention means rotating a slightly bent nail axle, or adding shims to the nail groove.

This PD wheel alignment is like laser alignment of people car wheels in the garage. The technicians do not take the car for a test drive to see if the people car rolls straight. Instead they just align each wheel separately.

Step 2: Orient front dominate wheel axle until car rolls straight
Gently roll the car along a smooth ground and watch if and how it turns. Find the front dominant wheel by pressing down on the front and seeing which wheel does not push down. Replace that dominant nail axle with a slightly bent nail and rotate that nail until the car rolls straight. Or add a shim to the groove.

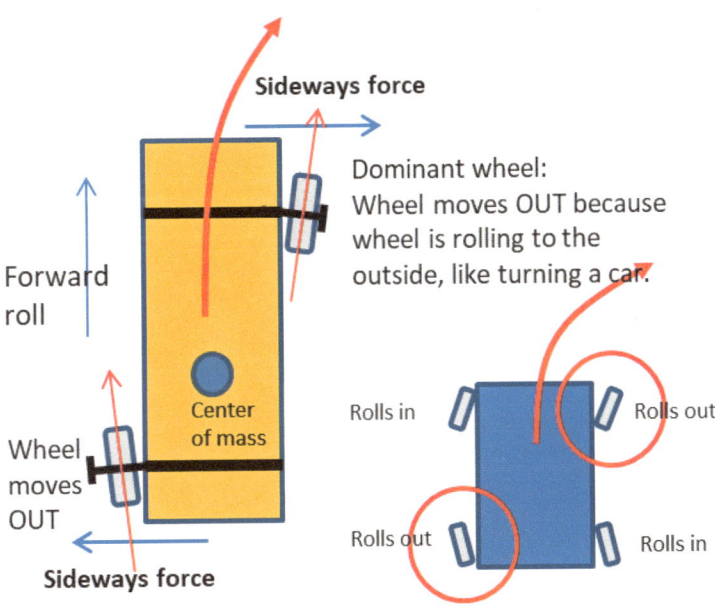

Summary Questions and Answers

> Who does the building, who makes the design, and how is gravity involved?

1. **Dad or Mom, are you going to do all the work?** No, when your kid is very young, they can make the concept and paint. When the kid gets a little older, they can cut the shape of the body, paint, and maybe align the wheels. The parent can make their own separate car.
   - No, if the scout is ready, Pinewood Derby is for your kid to learn about saws, accomplishment, and friendly competition. Dad or Mom can build their own second car side-by-side with the scout.
   - If the scout is not ready, then any and all help is part of the learning and bonding.

2. **What color paint goes faster? Red or White?** Trick question, the color does not matter. Maybe the smoothness matters, but not the color.
   - Neither, it doesn't matter.

3. **What's the most important thing to get right to get a fast car?** Align the wheels to go straight and avoid monorail drag.
   - Okay, the answer is 'align wheels to go straight and avoid rubbing against monorail' …the main two points of book. If the car is rubbing against the monorail, that is like riding the brakes. The next important things are smooth nail axles and graphite lubricant.

4. **How early do you need to plan to make a Pinewood Derby car?** Not very early, the project is about fun.
   - Planning could be a week, gathering paints and saws and brainstorming car ideas, or it could be a ½ hour before the races with the unpainted block car, as is, with a marker that writes 'winging it'. Either way, the car is fun.

5. **How is the Pinewood Derby car like a People car?** The PD car is mostly a scaled down people car shape.
   - A Pinewood Derby car has about the same wheel base ratio, and more than highway speeds if scaled up to the size of People cars.

6. **How are Pinewood Derby cars not like People cars?** The PD car can not turn and does not have an engine.
   - The Pinewood Derby car has no engine to go up hill or to drive around on flat roads, and the car can't turn. The Pinewood Derby car crashes into a cushion at the end of every race and survives to do many more races. In the People car world, that is a 'totaled car' collision. You'd be lucky to walk away.

7. **Does gravity help even People cars go down a hill, like a Pinewood Derby car?** Yes, you may have noticed in a People car that the car can accelerate down hill.
   - Yes, the engine can idle and not burn much gas coasting down a hill. For an electric People car, the energy of coasting down the hill can be used to re-charge the battery, with any hybrid combustion engine turned off.

8. **How efficient is a Pinewood Derby car?** A PD car is very efficient when designed without monorail drag. People cars only use about 30% of the gasoline energy to move forward. PD cars use all of gravitational energy to roll down hill.
   - A Pinewood Derby car can be very efficient, with the right tuning to reduce the drag forces - no monorail rubbing or braking, good lubricant on nail axles, and low profile for low air drag. These Pinewood Derby cars can get to 80% efficient. Unfortunately, this PD car does not have an engine, so you won't be driving a Pinewood Derby car along flats and up hills to the grocery store anytime soon. A falling rock can be 100% efficient before air drag starts to slow it down, but you don't drive it because a rock is only good at going down hill.

Appendix A: Wheel Alignment Process with Shims or Nail Rotation

Here are specific steps to align your wheels, in a two step process.

"Let's get rid of the frustrating, random, trial and error for wheel alignment, and get a system. First, look at wheel movement. Second, look at car movement. You can use shims, tap and compress the wood groove, or rotate a slightly bent nail, to get nails straight. The goal is for wheels to not move in or out as car is rolling forward. Moving in or out means the axle is bent."

First, roll car forward and look at the wheels.

See if you can tell if any of the wheels are moving IN or OUT while rolling forward. If so, then shim, re-seat the wood by tapping nail with hammer, or rotate or bend the nail to get the wheel more stable.

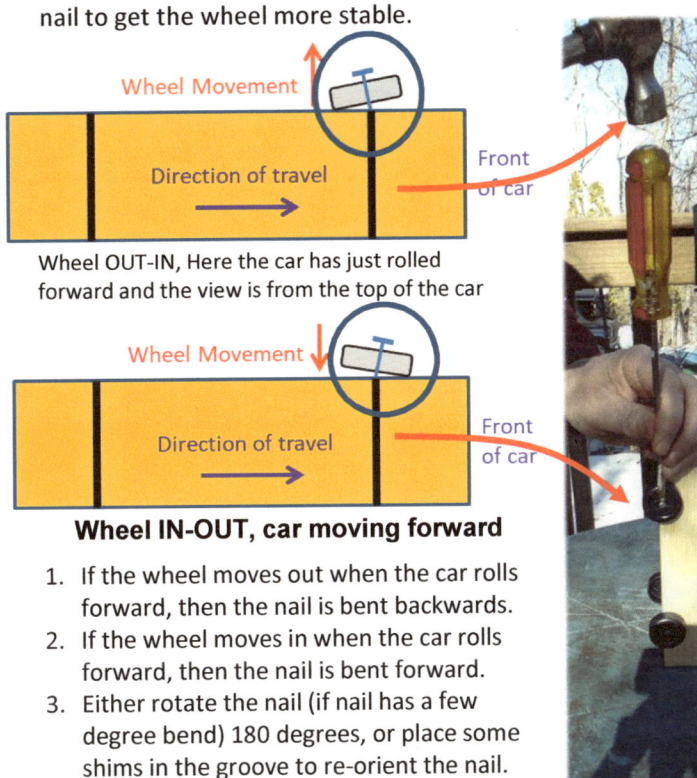

Wheel OUT-IN, Here the car has just rolled forward and the view is from the top of the car

Wheel IN-OUT, car moving forward

1. If the wheel moves out when the car rolls forward, then the nail is bent backwards.
2. If the wheel moves in when the car rolls forward, then the nail is bent forward.
3. Either rotate the nail (if nail has a few degree bend) 180 degrees, or place some shims in the groove to re-orient the nail.

Use shim, or rotate nail if bent, to get wheel to not move in or out as roll car.

Slightly tap nail with a hammer to straighten, compressing wood on one side.

Second, let the car coast forward and look at the turn.

Just like a regular people car, just think how wheels would be turned on a people car, mainly on the front tires, when the people car is turning. Then bend or rotate the nail the other way, to straighten out the coasting.

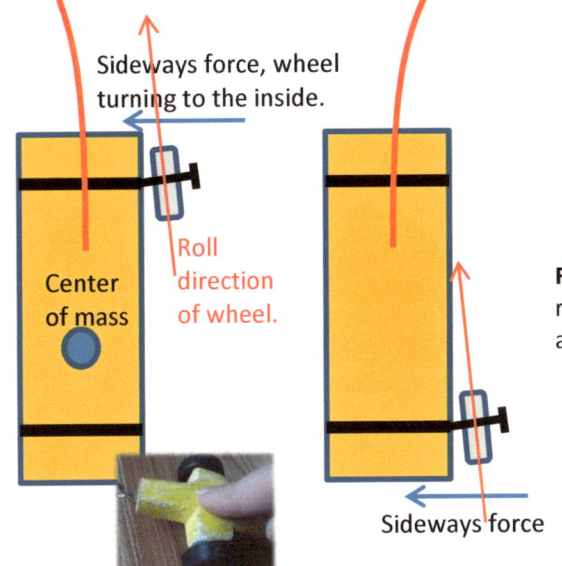

Find dominant wheel

1. Find dominant front wheel by pushing down on the front. The side that tips down is the minor wheel, and the side that stays firm is the dominant wheel.
2. Rotate the nail of the dominant wheel until the car goes straight, or the wheel does not push in or out as the car is rolled forward and backward.
3. You can tell how the nail is angled by imagining a people car turning the front wheels. The nail is bent in that direction.
4. Rear tires turn in other direction because the rear wheel is behind the center of mass, and the sideways force is operating behind this center of mass.

Four wheel turn: Above is a prototype regular car, if could steer all wheels to do a quick turn.

Wheels need to be aligned

Disney World monorail

Roller roaster under monorail

Trick 1 🚦 Drive Straight

Look at wheel IN-OUT sideways motion and car turning, and apply shims or bend a nail.

Appendix B: Definitions of Engineering and Physics Terms

> These engineering terms are used in any prediction of motion.

Force: A push or pull on something, like gravity on the Pinewood Derby car, or lifting the car and bringing it back to the starting gate.

Potential energy: Gravity energy that can be converted to Kinetic energy, like being up high. It is the same as the work energy that someone had to put into the object to lift it to the top of the race track.

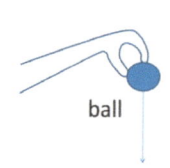
ball

Height for Gravitational Potential Energy

Kinetic energy: Energy of movement. You had the height, now you have the speed. Look for a velocity squared in the expression for energy of movement.

Energy of Motion
Convert gravity energy to motion energy

Rotational or Spin energy: Energy of spinning something. A toy top has a lot of rotational energy, even though it is not moving forward. Look for spin frequency (an angular velocity) squared in the expression for energy.

Energy in spinning tire
This spin energy steals some of the motion forward energy.

Momentum: People say that what's in motion wants to stay in motion, well that is momentum, and the expression for it is momentum = mass * velocity. A truck coming at you wants to stay in motion more than a VW Beetle or a motorcycle. An object with more momentum has more energy and takes more force to stop.

Oh NO!

Trucks and falling pianos have huge momentum
Don't stand in front or underneath them.

Conservation of Momentum: When two objects collide, the final momentum after collision is always the same as the sum of the two momentums before the collision, even if most of the collision energy goes up in heat.

- On a pool table, the collision of a first moving ball with a second non-moving ball of equal weight will stop the first moving ball and the second non-moving ball will move at the same speed as the first ball: momentum is conserved.
- On a pool table, two equal balls come directly at each other at equal and opposite velocities, for a starting net momentum of zero. They hit and both stop dead, still with a total momentum of zero: momentum is conserved even though all the energy went into heat.
- On Newton's cradle, two balls hitting from one side will cause the two balls on the opposite side to move.

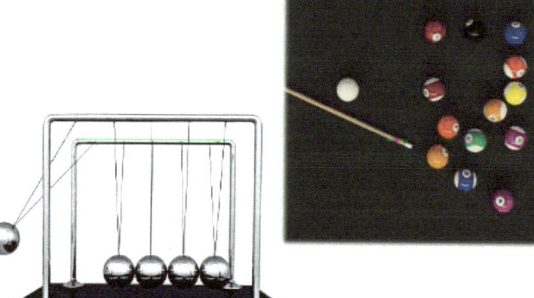
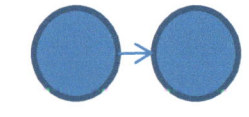

Pool table, with collisions and energy transfer.

Newton's cradle, demonstrating conservation of momentum.

Maybe some of your hands-on tinkering will help you speak the lingo, like Energy and Momentum.

Appendix C: 'Street Legal' Race Rules, Rules, Rules ...

Gotta have rules to make a more even playing field...

At weigh-in or registration before the races, the rules are checked. You cannot change your car after weigh-in. The car is taken away from you and stored in a box until the races, so be sure to add lots of graphite lubricant just before weigh-in.

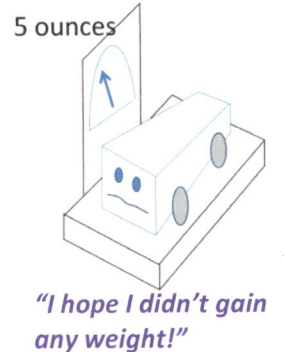

"I hope I didn't gain any weight!"

The rules for the pinewood cars exist to make the races competitive, and they are much less complex than the rules for a Formula 1 race (horsepower limits, weight limits). These pinewood rules make the fundamentals so important: 1) make that car go straight, so learn about aligning each wheel, 2) get nail friction down, so get graphite, 3) get a low profile to reduce air drag, and 4) get weights in back.

At some point, you might want to throw up your hands and say there is not much more to optimize because the rules restrict engines, CO2 rocket thrust, or bearings. Of course, this frustration just ignores the fact that you must perfect the fundamentals: Don't rail against the rules about no bearings when there is plenty to optimize on these simple cars. That said, you can always optimize a fun car too, like a banana car.

There is some logic behind some Pinewood Derby rules:

Gravity Power:
- Cannot have an engine to power the car, or a wound up spring

5 ounces weight limit:
- Air Drag: Heavier cars have an advantage because air drag depends on shape and not the weight of the car. Heavier cars will just plow right through the air drag (more energy is in the Kinetic Energy, forward movement, of the car). On the other hand, other drag sources won't improve with extra weight, like axle and monorail drag: these frictional forces with the track should all be proportional to the weight of the car, so no advantage there.
- Heavier cars will allow more energy in the forward motion, instead of energy in the rotation of the wheels, which stay the same using the same plastic wheels.
- You can buy little lead weights that stick to the car, to just get the weight up to 5 ounces at the registration for the car. Or drill a little of the lead or wood weight away if you are over 5 ounces.

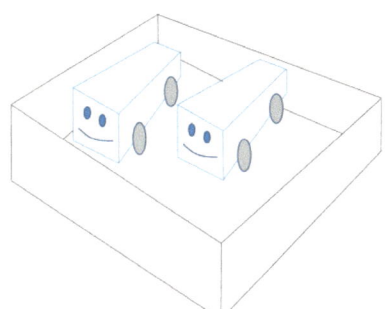

"Into the untouchable box they go, after check-in."

Height limit:
- Taller cars just do not fit under the finish banner. Taller cars, like high center of mass SUVs, can also tip over and damage other cars. For speed's sake, let's not forget that tall cars also have more air drag.

Axle separation limit:
- Moving the axles farther apart is not allowed. The distance between axles must be between 4.0 and 4.5 inches.
- Farther apart means that the car will travel straighter without turning, which makes less frictional drag against the monorail.

No loose parts:
- Nothing can fall off the car during the race.

Most racing sports, with people or not with people, are no stranger to rules, however. Rules make the race more even and competitive, where human skill is needed instead of unlimited horsepower. Look at NASCAR, with engine size limits, weight limits, wheel base limits, where the winning car is taken apart to confirm it followed the rules. Look at bicycle endurance races (Tour de France), with no-drug rules and no hidden electric batteries and motors in the bicycle frame.

Mostly the rules for Pinewood Derby boil down to no engine and a 5 ounce weight limit, along with length and height rules.

'Street Legal' Rules, continued

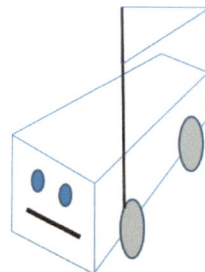

"The official rules: the not-to-be-questioned made-in-stone rules."

> Here are all the rules for a Pinewood Derby car:

Cub Scout Grand Prix Pinewood Derby Guidebook

1. The car must have been made during the current year (the year in which the derby is held).
    - *So you can't use 'ringers' from last year.*
2. The width of the car shall not exceed 2-1/4 inches.
    - *So you won't hit neighbors cars.*
3. The length of the car shall not exceed 7 inches.
    - *So cars have equal gravity energy and turning behavior.*
4. The wheel base shall be between 4.0 and 4.5 inches.
    - *Imposes equal turning behavior even when wheels are nearly perfectly aligned.*
5. The weight of the car shall not exceed 5 ounces.
    - *Imposes equal best-case slow down from air drag losses.*
6. Axles, wheels, and body wood shall be as provided in the kit.
    - *So have simple and do-able design.*
7. Wheel bearings, washers, and bushings are prohibited.
    - *The car stays within a younger scouts abilities.*
8. The car shall not ride on any kind or type of springs.
9. Any details added must be within length, width, and weight limits.
10. The car must be freewheeling, with no starting devices.
11. No loose materials of any kind (such as lead shot) are allowed in the car.
    - *So the car is safe and does not use rockets, or shoot anything backwards.*
12. The official number must be clearly marked or visible on both sides of the car.

https://en.wikibooks.org/wiki/How_To_Build_a_Pinewood_Derby_Car/Rules

Rules are everywhere in sports to make an even competitive playing field. Some example rules are:

School sports, like soccer or basketball:
- Athletes must compete only with kids of same age, or be within a narrow age group.
- There are huge experience differences and muscle changes in pre-teen and teenage early years.

Golf:
- Golf ball manufacturers can't make a golf ball that goes too far (the golf association gives distance limits, which control dimples and springiness of inside of ball).
- We want the golfer's swing to control the distance, not a better ball.

Swimming:
- Swimmers can't use a full body wetsuit, which could provide more of a flotation advantage.

Boxing and Wrestling:
- Athletes must fight within their weight group.
- Heavier weight groups have more muscle and longer arm reach

Baseball:
- No corking the bat. Strange rule, because hollowing the bat tip means the batter can swing the bat faster and hit the ball more easily, but the smaller bat mass means the ball does not go as far.
- Can't give the pitcher a break of an inning, to test endurance and to stop different pitchers for left and right handed batters.

Skiing:
- Skiers can't use aerodynamic suit, with envelope like a wing.
- The skier's crouch or tuck should control whether they win.

> **Just because, at your local pack races, you are casually allowed a longer wheel base or loose parts in the pack races, doesn't mean you're allowed it at the regional level.**

Index

Cars for people
- Air Drag — 52-57
- Bearings — 33, 47-49, 51
- Engine efficiency — 29
- Painting — 14
- Racing, Nascar — 18, 20
- Wheel energy — 62-66

CO_2 Car — 77-79

Drag, Monorail
- Estimate of car lengths lost — 30
- Ways to get car to drive straight — 36-44, 93
- Wood versus metal monorail drag — 39-40

Drag, Nail axle
- Comparison to bearings — 33, 47-48
- Estimate of car length lost — 30
- Ways to reduce nail axle drag — 47-51

Drag, Air
- Coefficient of drag — 52-57
- Estimate of car length lost — 30

Energy, Kinetic (motion)
- Car — 61, 90, 94
- Wheels — 62-66, 90, 94

Energy, Potential gravity energy of car — 29, 58-61, 90

Gravity for Pinewood Derby — 4

Gravity, common behavior and uses — 4-5

History of Pinewood Derby — 6-9

Monorail
- Concept — 36-39
- Drag — 36-40

No-rules cars:
- Propeller car — 80, 84-85
- Rocket car — 80, 81-83
- Throwback car — 80, 86-87

Painting — 14

Race Schedules — 21

Race rules Pinewood Derby — 17-18

Rules for real cars
- Indy car — 18
- Nascar — 18
- Street cars — 18

Slower speed techniques — 28, 34-35

Track shape — 4, 72-74

Weight in back — 58-61

Wheel alignment — 41-44, 93

Wheelbase — 45-46

Back Cover

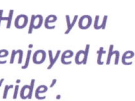
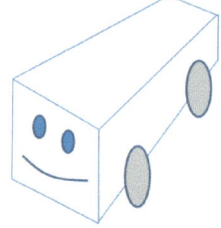

Hope you enjoyed the 'ride'.

Court Rossman first got into building Pinewood Derby cars when helping his son and daughter with Cub Scouts. Pinewood derby cars allow kids to use wood, hammers, nails, and creativity and competition. The kids can build the car themselves, although it is possible that the scout leaders themselves learn more from doing the activity.

This book compares the Pinewood Derby cars to real cars, in a fun and informing way, and could be read by both the parent and the kid together, to trigger and answer questions. The general concepts are all in the main part of the book. Part of the emphasis of the this book is a comparison of PD cars to people cars, and part is a description of some of the physical reasons a PD car goes fast, avoiding drag. This book is good in combination with other Pinewood Derby books that more directly describe a step by step process to making a good Pinewood Derby car.

Court is participating in the emphasis on hands-on learning. Science kits are readily available these days, and that is great. He just wants to help that trend, so the next generation has practical knowledge and creativity.

Court has also published 'Gamut of Speedy Rockets', 'Water Bottle Rockets', and 'Magnets, Motors, and Generators'. Court has a life-long interest in physics and science, and has a Ph.D. in Physics. He volunteered as an assistant scout master for Boy Scouts.

Thanks to the Cub Scout Pack 525 for running some exciting Pinewood Derby races, and inspiring this book. Cub Master Mark Collins ran the races. Thanks to Jane Henderson for suggesting 'Dad, Sir Isaac Newton, and Me' in the title. Thanks to Troup 15 for the CO_2 car. Thanks to people who posted pictures on the internet, many of which are used in this book.

References:
http://www.pinewoodprofessor.com/friction.html

http://en.wikibooks.org/wiki/How_To_Build_a_Pinewood_Derby_Car/Assembly

http://www.winderby.com/m02_040829.html

http://science.howstuffworks.com/transport/engines-equipment/bearing1.htm

https://en.wikibooks.org/wiki/How_To_Build_a_Pinewood_Derby_Car/Rules

The Pinewood Derby showcase in this book

www.ingramcontent.com/pod-product-compliance
Lightning Source LLC
Chambersburg PA
CBHW041325290426

44109CB00004B/125